FIBER REINFORCED COMPOSITES

FIBER REINFORCED COMPOSITES

By

Dr. Virendra Dedha

2016

SBS Publishers & Distributors Pvt. Ltd.
New Delhi

ISBN 13 : 9789380090740

First Published in 2016

Published by:

SBS PUBLISHERS & DISTRIBUTORS PVT. LTD.

2/9, Ground Floor, Ansari Road, Darya Ganj,

New Delhi - 110002,

INDIA

Tel: 0091.11.23289119 / 41563911

Email: mail@sbspublishers.com

www.sbspublishers.com

Preface

A fiber-reinforced composite (FRC) is a composite building material that consists of three components--the fibers as the discontinuous or dispersed phase, the matrix as the continuous phase, and the fine interphase region, also known as the interface. This is a type of advanced composite group, which makes use of rice husk, rice hull, and plastic as ingredients. This technology involves a method of refining, blending, and compounding natural fibers from cellulosic waste streams to form a high-strength fiber composite material in a polymer matrix. The designated waste or base raw materials used in this instance are those of waste thermoplastics and various categories of cellulosic waste including rice husk and saw dust. The FRC is high-performance fiber composite achieved and made possible by cross-linking cellulosic fiber molecules with resins in the FRC material matrix through a proprietary molecular re-engineering process, yielding a product of exceptional structural properties. Common fiber reinforced composites are composed of fibers and a matrix. Fibers are the reinforcement and the main source of strength while the matrix 'glues' all the fibers together in shape and transfers stresses between the reinforcing fibers. Sometimes, fillers or modifiers might be added to smooth manufacturing process, impart special properties, and/or reduce product cost. The FRP composite manufacturing can be an energy-intensive process with high heat and pressure needed to bond the composite material together.

Editor

Contents

Chapter 1

STRATEGY FOR INTRODUCING 3D FIBER REINFORCED COMPOSITES WEAVING TECHNOLOGY

Owais Anwar Golra[a*], Jawad Tariq[b], Nadeem Ehsanc an[d] Ebtisam Mirz[ad]

[a,b,c,d]Center for Advanced Studies in Engineering, Ataturk Avenue, G-5/1, Islamabad, Pakistan

INTRODUCTION

Textile plays a fundamental role in meeting man's basic needs. Initially, textiles were used for wearing and decoration purposes only but now textiles and clothing are truly global industries and therefore play a vital role in modern economic development. In last 50 years, a drastic change has come in the application of textiles and therefore it has attracted the attention of scientists and technologists to do more and more research in this industry.

Moreover, at the same time a new concept has come up in the field of textile, known as "technical textiles" due to its interesting properties and new fields of applications. In order to remain competitive in the changing textile market trends, it is necessary for a country to adopt new emerging technologies as well as new textile designing and manufacturing techniques.

General definition of textile refers to production of any product from fibers. Various techniques are being used for manufacturing fabric/product from fibers, such as, weaving, knitting, stitching and braiding. Weaving is widely used as a process for single layer and broadcloth fabric manufacturing within the textile industry, but knitting, felting, net making and non-woven processes or a combination of these also exist in the industry. Three dimensional woven technical textile structures are developed and produced by using advanced manufacturing techniques like 3D weaving, 3D braiding, 3D knitting and 3D stitching etc. This paper reviews the concept of 3D fiber reinforced composites weaving technology and proposes a strategy for introducing the technology.

Literature Review

Nowadays, the conventional textile is no more a profitable business even after huge investments, due to high competitive market and in such kind of business it is nearly impossible to sustain remarkable profitability for a longer period of time [2], particularly in those developing countries where production cost is increasing continuously due to various reasons. So there is a dire need of technological revolution for economic development of such countries. It is also believed that the intelligent application of new technologies in developing countries like Indonesia, India, Brazil and Pakistan etc. can certainly boost up their processes of economic growth [8] but first of all they have to develop their strong SME sector because this sector has the ability to respond quickly to technological changes and most of the developed countries manufacture these products under the SME sector [2].

The application of textile material in technical textiles has given a thrust to fiber technology so as to realize the advancement in needs of the society. Technical textile materials and products manufactured are quite different from non-technical textiles because of their technical and performance properties rather than their aesthetic or decorative characteristics [2], [4]. Few of the application of technical textiles are clothing, medical and health care products, automotive components, aircraft, shuttle and satellite components, building material, Geo-textiles, agriculture, sport and leisurewear, filter media, environmental protection, etc. [Source: Technitex 2006].

World consumption of technical textiles by product types is given in Table-1; World consumption of technical textile by region is given in Table-2 and World consumption of technical textile by application is given in Table-3.

Table.1 World Wide Consumption of Technical Textiles (By Products), Quantity: tonnes, Value: US $ Million

Product	Quantity		Growth % per Annum	Value		Growth % per Annum
	2000	2005		2000	2005	
Fabrics	3,760	4,100	1.7	26,710	29,870	2.2
Non-woven	3,333	4,300	5.4	14,640	19,250	5.6
Composites	1,970	2,580	5.5	5,960	9,160	5.6
Other Textiles	7,687	8,703	3.4	12,950	14,060	3.3
Total	16,750	19,683	3.9	60,260	72,340	3.7

Source: David Rigby Associates/ Techtextil

Table 1 [4] shows that there is a continuous annual growth in consumption of technical textiles by product i.e.fabrics, non-woven, composites and other textiles. There is not much increase in the consumption of fabric but still there is increased by 1.7% followed by non-woven products which are increased by 5.4%. The increase in consumption of composites is highest as compared to all other products which are reported to be 5.5% from the year 2000 to 2005. In the end other textiles and total products are increased by almost 3.4 and 3.9% respectively from year 2000 to 2005. Along with the increase in growth %age of technical textile, the value of product in US $ Million has also increased so there is good investment opportunities for new countries

Table 2. World Wide Consumption of Technical Textiles (By Region), Quantity in 000 tonnes

Region	1995	2000	2005	2010
North America	3,584	4,184	4,774	5,591
South America	705	847	1,004	1,230
West Europe	3,002	3,614	4,107	4,760
East Europe	493	584	666	817
Asia Excl-China	3,895	4,449	5,220	6,348
China	1,515	2,155	2,871	3,808
Other Countries	778	917	1,041	1,219
Total	13,972	16,750	19,683	23,773

Source: Technical Textiles and Industrial Nonwovens: World Market forecasts to 2010, David Rigby Associates, 2002

Table 2 [4] demonstrates the actual and expected annual consumption of technical textile by region for the year 1995, 2000, 2005 and 2010. Asia, North America, West Europe and China are among the leading consumers of technical textile products by region. From the above table it is clear that the consumer market has shown a good trend in consumption of technical textile and which is expected to increase more in future.

Table 3 World Wide Consumption of Technical Textiles (By Application), Quantity: 000 tonnes, Value: US $ Million

	2000		2005	
Area of Application	Quantity	Value	Quantity	Value
Transport Textiles (Auto, Marine, Aero)	2,220	13,080	2,480	14,370
Industrial Products & Components	1,880	9,290	2,340	11,560
Medical & Hygiene Textiles	1,380	7,290	1,650	9,530
Home Textiles, Domestic Equipments	1,800	7,780	2,260	9,680
Clothing Components (Thread, Interlinings)	730	6,800	820	7,640
Agriculture, Horticulture & Fishing	900	4,260	1,020	4,940
Construction Building & Roofing	1,030	3,390	1,270	4,320
Packing and Contamination	530	2,320	660	2,920
Sports & Leisure (Excluding Apparel)	310	2,030	390	2,210
Geo Textiles, Civil Engineering	400	1,860	570	2,360
Protective & Safety Clothing	160	1,640	220	2,230
All Others	5,410	520	6,003	3,580
Total	16,750	60,260	19,683	75,340

Source: Technical Textiles and Industrial Nonwovens: David Rigby Associates, 2002

Table 3 [4] shows the worldwide annual consumption of technical textiles by application in different industries, total of which is increased by almost 3% from the year 2000 to 2005. A continuous increase in the demand for technical textile in each of the industry has been observed which means that scope of the application of technical textiles will increase in the future.

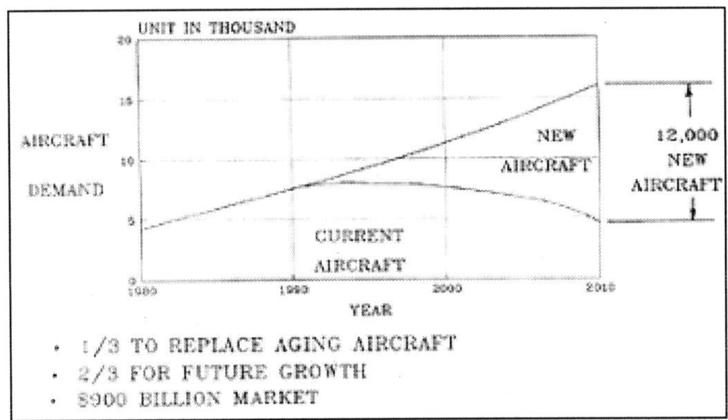

Figure.1. Commercial Aircraft Demand by the Year 2010

Figure. 1 shows the expected new world market demand of the aircraft, which is expected to be almost 12000 new aircraft by the year 2010 [National Aerospace Laboratory (NAL), 6-13-1, Ohsawa, Mitaka, Tokyo 181, Japan, December 7, 1992]. Aircraft is one of the industries which have vast application of technical textiles (composites). So the increase in aircraft demand means that there would also be increase in the demand of composites usage in future. From the above tables and fig, it is understandable for a country to seek technological revolution in the field of textile composites particularly 3D fiber reinforced composites because of their high tech nature for competitive position.

Figure 2 [10] illustrates the hierarchy of textile fiber reinforced composite production. The first step is the formation of yarns from fibers. In the second step, the yarns are woven into fabric structure. The fabrics are then laid up into required shape and then stitched together to create a structured preform. Finally, through consolidation process composite part is infiltrated by resin and then curing in a mould.

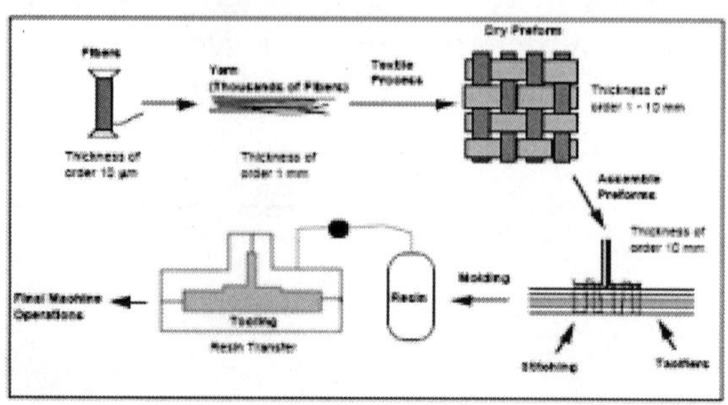

Figure.2. Shows the basic steps involve in the production of textile fiber reinforced composite structure.

Fibers embedded in a rigid polymer matrix are termed as composite material [1]. Different types of fibers are being used as reinforcement material such as carbon fiber, fiber glass, Kevlar, boron and spectra etc. Initially 2D fiber reinforced composites were used for various industrial application but due to de-lamination problem and shortage of 2D laminates along with several other reasons, 3D fiber reinforced composite material were developed in late 1960s. Braiding was the first method used for 3D fiber reinforced composite manufacturing but it was not enough to meet the increasing demand of various industries, therefore from 1985 to 1997 "Advance Composite Technology Program (ATCP)" was launched and as a result of which various other techniques were invented such as 3D weaving, 3D knitting, and 3D stitching [1].

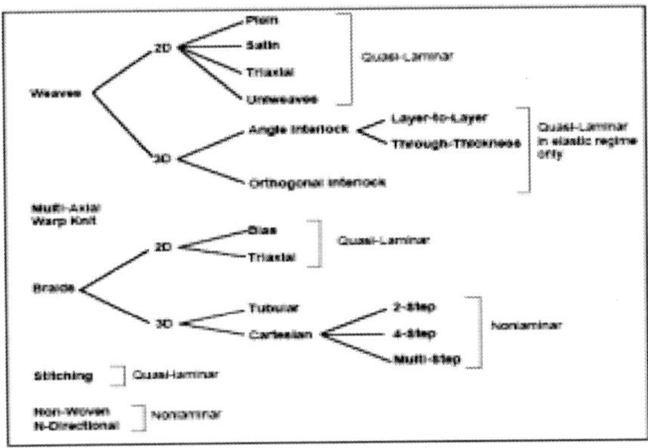

Figure.3. Categories of Textile Composites for High end Applications

Figure 3 [10] introduces the most important groups of textile composites and manufacturing techniques. Column wise from left to right, first textiles processes are categorized followed by the dimensional structure and manufacturing technique/design of the textile perform. At the end differentiate each category of 2D and 3D textile composite according to their macroscopic properties. Textile composites which do not behave as laminates and have equal load bearing capacity along the three axes are known as non-laminar and composites which behave as laminates and majority of fiber lie in plane are called as quasi-laminar [10].

Previously 3D woven performs were only manufactured in simple structures but now advance techniques are also invented [6]. Although various techniques have been developed for the manufacturing of 3D woven performs, or fabrics [6] and these performs are also considered among the technical textiles because of their high end use application like automotive, marine, civil, medical and aircraft industry [5]. Eventually, there are still very few countries which can produce such structures or which have the knowledge of these advance manufacturing techniques, so there is a dire need of transfer of these technologies to other part of the world to meet the increasing demand of the industries.

3D WEAVING

Weaving is simply producing a fabric by the interlacing of two sets of yarns i.e. warp and weft. The first major difference between conventional weaving and 3D weaving is the need to have multiple layers of warp yams to achieve the required thickness of the fabric [1]. Fig. 4 shows the view of 3D weaving of carbon fiber. Moreover 3D weaving process involves double directional shedding technique which require two sets of weft yarn one for horizontal and the other for vertical filling, a multiple layers of warp yarns and warp shed formation both row wise and column wise [6].

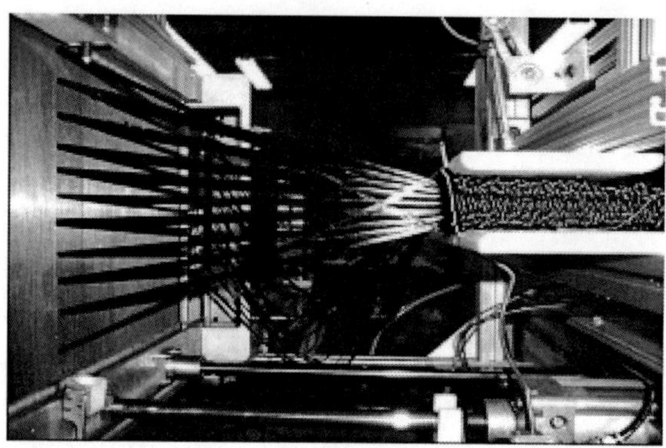

Figure.4. Three dimensional (3-D) weaving

An advantage of 3D weaving is that 3D fabrics can also be manufactured on standard industrial loom with a little modification [3], [5]. Conventional technologies like Dobby and Jacquard are the examples of standard industrial looms which can be used for manufacturing 3D woven performs [7], thereby, widening the scope of conventional technologies. There are different methods of 3D fabric formation i.e. non-interlacing, interlacing type, multi axial and double directional shedding etc [6] [7]. This paper reviews the concept of 3D weaving only and does not include the other 3D fabric manufacturing technologies.

ANALYSIS & DISCUSSION

Generally, the key need of this paper is the development of a robust strategy for introducing 3D fiber reinforced composite weaving technology to the areas of world which are lagging behind in this technology and thereby missing the opportunity for their technological advancement and economic growth. And also for countries which are already involved in composites weaving to enhance their capabilities to remain competitive in the global market.

The business environment is increasingly influenced by the pace of technological development and globalization of the world economy, which means that customers, competitors, suppliers and collaborators may be located anywhere.

Thus any region can take initiative to become supplier, competitor or collaborator. In consequence the requirement for companies to constantly strengthen their capabilities and find innovative ways becomes necessary for survival.

At the same time society faces numerous challenges, many of which also present an opportunity not only for companies but also for countries.

The global composites market has grown considerably in the last 10 years with current annual revenue for finished parts in excess of £2 billion. This is set to rise by another 25% in the next five years with the aerospace and advanced transport sectors. However, increasing economic growth is being achieved by regions with lower labor rates than the West [Ken Wappat, Chairman, NCN (National Composites Network) UK, April 2006].

To vision of the strategy is to maintain and grow an industry, with visionary research and technology development capability, delivered through a globally integrated academic/industry partnership

The level of skills in the industry is high and therefore there is a need to improve advanced skills to enter and retain competitive in the industry. The more open the country is in its trade policy, more R & D culture is imported and therefore more skilled is their labor force and more its trade with the developed countries but to adapt foreign technology openness only is not enough, also there is need to learn how to apply the new technology and then improve it gradually overtime [9]. Globalization of the composite industry has been moving forward at a significant pace with the development of worldwide partnerships by the major producers for integrated solutions and indigenous growth of low-labor cost markets.

Recent academic research has proven that engaging in overseas trade stimulates competitiveness and that those companies that are internationally active are generally more successful. Regional supply chain companies tend not to seek global business, but seek to win business from mainly regional sources of work.

Private Venture finance companies must show positive signs of targeting composite weaving as an investment opportunity. For the supply chain this may prove an attractive source of growth funding.

Steps to be taken:

To introduce 3D Fiber Reinforced Composites weaving technology in a country, following steps are to be taken;

- To identify major players with their respective market share
- To develop understanding with major clients i.e. Defence Forces, Textile Industry and Automotive Industry etc.
- To get hold of their requirements with technical data
- To develop indigenous team of engineers/technicians and send them abroad for research, development and training.
- To concentrate on major manufacturing organizations and to convince them to invest in 3D Weaving Technology.

CONCLUSION

It is a highly strategic industry with very low competition amongst the suppliers but at the same time also a very technical and advanced industry. For developing countries such technical infrastructure are expensive and need expertise and skill levels to use the benefits of such a technology by finding opportunities to export the product. Also one has to penetrate in this market by identifying a niche and gradually establishing itself.

REFERENCES

1. L. Tong, A. P. Mouritz,, and M. K. Bannister, "3D fiber reinforced composites" Elsevier science limited, 2002 {Chapter 1}.

2. "How to catch up with latest textile market trends," Pakistan, Dawn News, July17, 2006.
3. S. S. Badawi "Development of Weaving Machine and 3D Woven Spacer Structures for Lightweight Composites Material" PhD Dissertation, Dresden, 9-07-2007
4. N. A. Memon, N. Zamaan, "Pakistan lags behind in Technical Textiles" Journal of Management and Social Sciences, Vol. 3, No. 2, (Fall 2007) 120-127.
5. A. P. Mouritz, M. K. Bannister, P. J. Falzon, K. H. Leong "Review of applications for advance three-dimensional fiber textile composites" ELSEVIER, Composites: Part A 30 (1999) 1445-1461.
6. N. Gokarneshan, P. Dhanapal, "3-D Woven Preforms: Review on its Manufacture, Technical Parameters, Characterization and End uses," Volume 88, February 2008.
7. J. A. Sodan, B. J. Hill, "Conventional Weaving of shaped performs for engineering composites." ELSEVIER, Composites part A 29A (1998) 757- 762.
8. U. Colombo, "The technical revolution and the future of the third world," IEEE Technology and Society Magazine, spring 1991.
9. B. M. Hoekman, K. E. Maskus, K. Saggi, "Transfer of technology to developing countries: unilateral and multilateral policy options," ELSEVIER, World development Vol. 33, No. 10, pp. 1587- 1602, 2005.
10. B. N. Cox and G. Flanagan, "Handbook of Analytical Methods of Textile Composite" NASA Contractor Report 4750, {Chapter 2}

Chapter 2

STUDIES ON INTERACTING BLENDS OF ACRYLATED EPOXY RESIN BASED POLY(ESTER-AMIDE)S AND VINYL ESTER RESIN

Pragnesh N. Dave[1*], Nikul N. Patel[2]

[1]Department of Chemistry, KSKV Kachchha University, Gujarat, India;
[2]Institute of Technology, Nirma University, Ahmedabad, India.

ABSTRACT

Epoxy resin based Unsaturated poly(ester-amide) resins (UPEAs) can be prepared by many methods but here these were prepared by reported method [1]. These UPEAs were then treated with acrylate chloride to afford acrylated UPEAs resin (i.e. AUPEAs). Interacting blends of equal proportional AUPEAs and vinyl ester epoxy (VE) resin were prepared. APEAs and AUPEAs were characterized by elemental analysis, molecular weight determined by vapor pressure ohmmeter and by IR spectral study and by thermogravimetry. The curing of interacting blends was monitored on differential scanning calorimeter (DSC). Based on DSC data in situ glass reinforced composites of the resultant blends have been prepared and characterized for mechanical, electrical and chemical properties. Unreinforced blends were characterized by thermogravimetry (TGA).

INTRODUCTION

Both the polyesters and polyamides discussed have par-ticular individual properties and applications. Hence many researchers have synthesized co-poly (ester-am- ide)s from different raw materials in order to obtain the properties and applications of the individual ones into one segment. Some of the poly (ester-amide)s synthe-sized by different researchers. The three polymer candi-dates namely epoxy resin, unsaturated poly ester and polyamides are most widely versatile industrial materials and have broad spectrum of characteristics for wide applications ranging from aerospace to micro electronics. They are also important as laminating resins, molding composites, fibers, films, surface coating resins, fiber cushion [3,4]. Particularly Polyamides material used in the form of fibers as especially thermoplastics of par-ticular used in engineering applications. The glass fiber reinforced nylon plastics are now of substantial importance due to rigidity and creep resistance. Polyamides are also used in fiber application, automotive industries, valve covers, coatings [5,6].

Merging of all three segments (*i.e.* epoxy, ester and amide) into saturated and unsaturated polymer chain has been recently reported from our Indian scientists [7-10]. Certain properties of resins may also be improved via interact of with the other unsaturated resin is another possibility. In order to improve certain properties of such reported USPEAs their blending with commercial vinyl ester epoxy resin is possible. While vinyl ester resin is versatile industrial resin today [11,12] Hence the present paper comprises studies of interacting blending of re-ported unsaturated poly(ester-amide) resin with vinyl ester (VE) resin. The glass fiber reinforced composites of these blends have been fabricated and characterized by chemical, mechanical and electrical properties. The whole work is scanned in Scheme 1.

MATERIALS AND METHODS

Materials

Commercially available epoxy resin, diglycidyl ether of bisphenol-A and vinyl ester epoxy resin was obtained from local market.

The specification of diglycidyl ether of bisphenol-A (DGEBA) are as follows:

+ H_2N—R—NH_2

Maleic Anhydride (2.0 Mole)

Aromatic Diamines (1.0 Mole) (a-c)

0-5°C

HOOC —CH =CH —CO—HN —R—NH —OC —CH =CH—COOH

Unsaturated Bisamic Acids (1a-c)

Diglycidylether of Bisphenol-A

Unsaturated Poly(Ester-Amide) Resin (UPEAs 2a-c)

Acrylol Chloride

Acrylated Unsaturated Poly(Ester-Amide) Resin (AUPEAs 3a-c)

0.05% Hydroquinone

Stirr well for ten minutes at 85°C

Stirrer 50% of AUPEAs and 50% of VE resin at 80°C for 1-hour

AUPEAs-VE Blends (4a-c)

Benzoyl Peroxide

Δ

Glass Fiber Reinforced Composites (5a-c)

Where, R= (a) 4,4'-methylenedianiline (b) 4,4'-oxydianiline (c) 4,4'-sulfonyldianiline

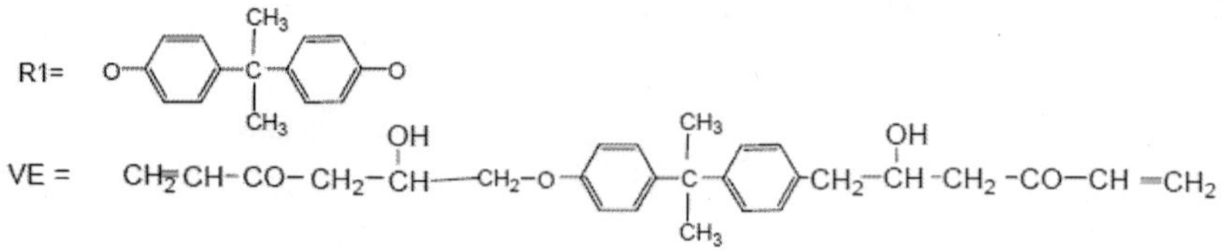

R1= O–C$_6$H$_4$–C(CH$_3$)$_2$–C$_6$H$_4$–O

VE = CH_2=CH–CO–CH_2–CH(OH)–CH_2–O–C$_6$H$_4$–C(CH$_3$)$_2$–C$_6$H$_4$–CH_2–CH(OH)–CH_2–CO–CH=CH_2

- Epoxy equivalent weight, 190.
- Viscosity 40 - 100 poise at 25°C.
- Density at 25°C, 1.16 - 1.17 g/cm.

Vinyl Ester Epoxy

The aromatic diamines used for the preparation of un-saturated poly(ester-amide) resin are,

- 4,4′-methylenedianiline
- 4,4′-oxydianiline
- 4,4′-sulfonyldianiline

Plain weave fibers, in the form of E-glass woven fab-ric (poly (ester-amide) compatible) 0.25 mm thick (Un-nati Chemicals, India) of a real weight 270 g m–2 were used for composite fabrication. All other chemicals used were of pure grade.

Synthesis of Unsaturated Bisamic Acids

The unsaturated bisamic acid was prepared by a simple addition reaction of maleic anhydride and diamines. These were prepared by using method reported in the literature [7,8]. The general procedure for the synthesis of unsaturated bisamic acid is as follows.

To a well-stirred solution of maleic anhydride (2.0 mole) in dry acetone, the solution of diamine (1.0 mole) in dry acetone was gradually added at 0°C -5°C within 30 minutes. After complete addition of the diamine, the re-action mixture was further stirred for half an hour at room temperature. The resulting unsaturated bisamic acid was then filtered, washed with dry acetone and air-dried. Unsaturated bisamic acid was obtained in the form of free flowing powder. The reaction scheme for the syn-thesis of unsaturated bisamic acid is shown in Scheme 1.

Synthesis of Unsaturated Poly(Ester-Amide) Resin and Acrylated Poly(Ester-Amide) Resin

The unsaturated poly(ester-amide) resin (UPEAs) was prepared by following the same method reported in [7,8] The general procedure is as follows:

Diglycidylether of Bisphenol-A (DGEBA) (1.0 mole) and unsaturated bisamic acid (1.0 mole) were charged in three necked flask equipped with a mechanical stirrer. The unsaturated bisamic acid was then treated with di-glycidylether of Bisphenol-A according to method re-ported for reaction of epoxy resin and carboxylic group [13]. To this 8.0% of the total weight of above, triethyl-amine (TEA) was added as a base catalyst. The reaction mixture was slowly heated up to 85°C with continuous stirring till the acid value fell below 60 mg KOH/gm. The resultant resin was then discharged and called unsaturated poly(ester-amide) resin (UPEAs) and their de-tails are furnished in **Table 1**. Further reaction of all these unsaturated poly(ester-amide) resin (UPEAs) was carried out with acryloyl chloride (*i.e.* acrylation) and the resultant products called acrylated poly(ester-amide)s (APEAs) and their details are furnished in **Table 2**.

Synthesis of Acrylated Poly(Ester-Amide) Resin and Vinyl Ester Resin Blend

When the acid value of acrylated poly(ester-amide)s fell below 55 mg KOH/gm, 0.05% of hydroquinone was added as an inhibitor. The whole reaction stirred well for ten minutes maintaining the temperature at 85°C. Then add 50% of APEAs and 50% of vinyl ester (VE) resin was added and stirred well at 80°C for one hour. The resultant APEAs-VE blends were obtained in the form of viscous syrup.

ANALYSIS AND THERMAL STUDY

Elemental Analysis

The C, H, N content of unsaturated poly(ester-amide)s (UPEAs) and acrylated poly(ester-amide)s (APEAs) was estimated by means of Thermofinagan 1101 flash ele- mental analyzer (Italy). The IR spectra were recorded in Kerr pellets on a Nicollet 760 D spectrometer. The number average weight of unsaturated poly(ester-amide)s (UPEAs) and acrylated poly(ester-amide)s (APEAs) was estimated by non-aqueous

Table 1. Characterization of UPEAs (2a-c)

UPEAs	Elemental analysis (Wt%) Calc. / (Found)			No. of –OH group per repeating unit	Number average molecular weight ($\overline{Mn}$) ± 60
	%C	%H	%N		
2a	65.62/64.23	5.72/5.50	6.43/5.83	1.96	3902
2b	64.06/63.83	5.20/4.86	3.64/3.15	1.82	3828
2c	60.29/60.12	4.90/4.63	3.43/3.05	1.81	4206

Table 2. Characterization of AUPEAs (3a-c)

AUPEAs	Elemental analysis (Wt%) Calc. / (Found)			No. of double bonds per repeating unit	Number average molecular weight ($\overline{Mn}$) ± 60
	%C	%H	%N		
3a	65.75/65.37	5.74/5.48	3.19/2.63	3.91	4133
3b	64.38/63.87	5.02/4.70	3.19/2.91	3.76	4135
3c	63.22/62.88	4.93/4.62	3.13/2.61	3.81	4428

Conduct metric titration fol-lowing by method reported in the literature [14]. Pyridine was used as a solvent and tetra-n-butyl ammonium hydroxide was used as a titrant.

Thermal Study

Curing of all APEAs-VE blends were carried out on a differential scanning calorimeter (DSC) by using benzoyl peroxide as a catalyst. A Du Pont 900 DSC was used for this study. The instrument was calibrated using standard indium metal with known heat of fusion

(ΔH = 28.45 J/g). Curing was carried out from 30°C - 300°C at 10°C min–1 heating rate. The sample weight used for this investiga-tion was in the range of 4 - 5 mg along with an empty reference cell. The results are furnished in **Table 3**.

Unreinforced cured samples of APEAs-VE blends were subjected to thermogravimetric analysis (TGA) on Du Pont 950 thermo gravimetric analyzer in air at a heating rate of 10°C min–1. The sample weight used for this investigation was in the range of 4 - 5 mg. The re-sults are furnished in **Table 4**.

COMPOSITE FABRICATION

The composites were prepared by using E-type of glass fiber. The glass fiber: APEAs-VE blend ratio is 60:40 (40% APEAs-VE blends). Suspensions of APEAs-VE blends were prepared in tetrahydrofuran (THF). In the above polymer suspension, 1% of ethylene dimethylacrylate (as a cross linking agent) with 0.05% benzoyl peroxide (as an initiator) were added and mixed well. The mixture was applied with a brush to a 200 mm × 200 mm glass cloth and the solvent was allowed to evaporate. The ten dried prepregs prepared in this way were then stacked one on top of another and pressed between steel plates coated with a "Teflon" film release sheet and compressed under 70 psi pressure. The prepregs stacks were cured by heating it in an autoclave oven at around 140°C for about 6 hour. The composites so obtained were cooled to 45°C - 50°C before the pressure was released.

Composite Characterizations

Chemical Resistance Test

The resistances against Chemicals of the composites were measured according to ASTM D 543. The results are furnished in **Table 5**.

Mechanical and Electrical Testing

- The Flexural strength was measured according to ASTM D 790.

- The Compressive strength was measured according to ASTM D 695.
- The Impact strength was measured according to ASTM D 256.
- The Rockwell hardness was measured according to ASTM D 785.
- The Electrical strength was measured according to ASTM D 149.
- The Tensile elongation was tested according to

Table 3 DSC Curing of UAPEAs-Vinyl ester epoxy resin Blends (4a-c).

UAPEAs-VE resin Blends	Curing Temperature (°C)		
	Ti	Tp	Tf
4a	130	151	172
4b	116	147	174
4c	126	138	186

Table 4. TGA of Unreinforced Cured Samples of UAPEAs-Vinyl ester epoxy resin Blends (5a-c)-BPO system

UAPEAs-VE resin Blends	% Weight loss at various temps. (°C) from TGA				
	150°C	300°C	450°C	600°C	750°C
5a	2.01	9.34	64.23	77.32	81.34
5b	1.87	9.21	68.56	76.63	79.34
5c	1.71	8.03	65.25	75.87	80.06

Table 5. Chemical, Mechanical and Electrical Properties of Composites Based on APEAs-MMA Blends

Composites	% Change on exposure to 25% (W/V) NaOH		Compressive strength (MPa)	Impact Strength (MPa)	Rockwell hardness (R)	Electrical strength (in air) (kV/mm)
	Thickness	Weight				
5a	0.78	1.07	460	424	111	22.65
5b	0.81	1.10	467	456	104	19.45
5c	0.79	1.02	478	447	106	23.35

All mechanical and electrical tests were performed using three specimens and their average results are sum-marized in **Table 5**.

RESULTS AND DISCUSSION

The unsaturated bisamic acids were prepared by the reac-tion of malelic anhydride (2.0 mole) and aliphatic dia-mines (1.0 mole) by following method reported in litera-ture [7,8]. Unsaturated poly(ester-amide)s (PEAs) was prepared by reaction of epoxy resin (DGEBA) with un-saturated bisamic acids using triethylamine (TEA) as a base catalyst. The post reactions of all these unsaturated poly(ester-amide)s (PEAs) were carried out with acryloyl chloride. The resultant products are called acrylated poly (ester-amide)s (APEAs). Blending of acrylated poly(es- ter-amide)s APEAs with vinyl ester (VE) resin was also carried out. The resultant products are called APEAs-VE blends.

The C, H, N content of all the unsaturated poly(ester- amide)s (PEAs) and acrylated poly (ester-amide)s (APEAs) were estimated by means of Thermofinagan 1101 Flash Elemental Analyzer (Italy). The values of C, H, N of each of the unsaturated poly(ester-amide)s (PEAs) and acrylated poly (ester-amide)s (APEAs) were consis-tent with their predicted structures and are furnished in **Tables 1** and **2**. The number average molecular weight of both unsaturated poly (ester-amide)s (PEAs) and acry-lated poly (ester-amide)s (APEAs) were estimated by non-aqueous conductometric titration method [14]. Their results are furnished in **Tables 1** and **2**, respectively. The results indicate that the degree of polymerization of both unsaturated poly (ester-amide)s (PEAs) and acrylated poly (ester-amide)s (APEAs) is about 6. The IR spectra were consistent with the ones expected from the struc-tures of the PEAs (3a-c) and APEAs (4a-c).

Numbers of hydroxyl group present per repeating unit in unsaturated poly(ester-amide)s (PEAs) was also ana-lyzed by employing acetylating method [15]. Also, acry-lated poly(ester-amide)s (APEAs) were characterized for the presence of double bonds per repeating unit employ-ing mercury-catalyzed bromate-bromide method [16]. Satisfactory results were found and the results are fur-nished in **Tables 1** and **2** respectively.

Curing of all these APEAs-VE blends were carried out on a differential scanning calorimeter (DSC) by using benzoyl peroxide as a catalyst. The data of DSC thermo-grams of all APEAs-VE are furnished in **Table 3**.

The unreinforced cured samples of APEAs-VE blends were also analyzed by thermo gravimetric analysis (TGA). The result reveals that the cured sample starts their degradation at about 150°C and their initial weight is about 2% - 3%. This small weight loss may be due to either in sufficient curing of components used or due to the catalyst used. A weight loss of about 10% - 11% is found at 300°C. However, the rate of decomposition in-creases very rapidly between 300°C to 450°C and the products are lost completely beyond 750°C. TGA data of all the samples are shown in **Table 4**.

The glass and carbon fiber reinforced composites of all APEAs-VE blends were prepared based on their DSC data. The composites were characterized for their chemi- cal, mechanical and electrical properties. Their results are furnished in **Table 5**. The results shows that composites have good chemical resistant property, good mechanical and electrical strength.

REFERENCES

1. V. Shukla, "Flow Modified Epoxy Resin: The Complete Solution of Aerosol in 2-Pack Epoxy Adhesive," *Pigment & Resin Technology*, Vol. 35, No. 6, 2006, pp. 353-357. doi:10.1108/03699420610711362
2. B. C. Samanta, T. Maity, S. Dalai and A. K. Banthia, "Influences of Amine-Terminated Oligomers on Glass Fibre-Epoxy Composite," *Pigment & Resin Technology*, Vol. 37, No. 1, 2008, pp. 3-8. doi:10.1108/03699420810839648
3. S. Khambete, "U. S. Patent," Vol. 7, November 2007, pp. 290-300.
4. M. Malik, V. Choudhary and I. K. Varma, "Current Statuse of Unsaturated Polyester Resin," *Journal of Macromolecular Science – Reviews in Macromolecular Chemistry & Physics*, Vol. 40, No. 2, 2000, pp. 139-165.
5. A. J. Pekarik, "U. S. Patent," Vol. 4, June 1989, pp 840,980.
6. S. Ahmad, S. M. Ashraf, A. Hasant, S. Yadav and A. Jamal, "Studies on Urethane-Modified Alumina-Filled Polyesteramide Anticorrosive Coatings Cured at Ambient Temperature," *Journal of Applied Polymer Science*, Vol. 82, No. 8, 21 November 2001, pp. 1855-1865.
7. H. S. Patel and K. K. Panchal, "Novel Unsaturated Poly-ester Resins Containing Epoxy Residues," *International Journal of Polymeric Materials*, Vol. 54, No. 1, 2005, pp. 1-7. doi:10.1080/00914030390224247
8. H. S. Patel and K. K. Panchal, "Flame Retardant Unsatu-rated Poly(Ester

Amide) Resins Based on Epoxy Resins," *International Journal of Polymeric Materials*, Vol. 54, No. 9, 2005, pp. 795-803. doi:10.1080/00914030490463133

9. H. S. Patel and B. K. Patel, "Novel Flame-Retardant Acrylated Poly(Ester-Amide) Resins Based on Bromi-nated Epoxy," *International Journal of Polymeric Mate-rials*, Vol. 56, No. 6, 2009, pp. 312-321. doi:10.1080/00914030902859257
10. H. S. Patel, and B. K. Patel, "Interacting Blends of Novel Acrylated Poly(Ester Amide)s Based on DGEBC with Styrene Monomer," *International Journal of Polymeric Materials*, Vol. 58, No. 12, 2009, pp. 654-664. doi:10.1080/00914030903146720
11. A. M. Atta, S. M. ElSaeed and R. M. Farag, "New Vinyl Ester Resins Based on Rosin for Coating Applications," *Reactive and Functional Polymer*, Vol. 66 No. 12, De-cember 2006, pp. 1596-1608.
12. I. K. Varma, "Matrix Resin for Composite," Department of Science and Technology Government of India, 1986, p. 148.
13. R. S. Darke, D. R. Egan and W. T. Murphy, "Elas-tomer-Modified Epoxy Resins in Coatings Applications," *ACS Symposium Series*, Vol. 221, 1982, pp. 1-20. doi:10.1021/bk-1983-0221.ch001
14. R. N. Patel and S. R. Patel, "Synthesis and Characteriza-tion of Poly(Keto-Amines). 1. Self Polycondensation of 4-Aminophenacyl Chloride," *Die Angewandte Makro-molekulare Chemie*, Vol. 96 No. 1, May 1981, pp. 85-92. doi:10.1002/apmc.1981.050960107
15. A. I. Vogel, "Quantitative Organic Analysis," CBS Pub-lishers, New Delhi, Vol. 3, No. 2, 1998, pp. 677-679.
16. A. I. Vogel, "Quantitative Organic Analysis," CBS Pub-lishers, New Delhi, Vol. 3 No. 2, 1998, pp. 765-766.

Chapter 3

COMPUTATIONAL MODELLING OF COMPOSITE MATERIALS REINFORCED BY GLASS FIBERS

Milan Žmindák[a*], Martin Dudinský[a]

[a]University of *Žilina,* Univerzitná 1, 010 26, Slovakia

ABSTRACT

Composite materials reinforced by micro particles are one of the topics of interest of researchers. Properties of fiber composites reinforced by long fibers significantly depend on the selection of fiber and matrix. By increasing length of fibers, the reinforcement is more effective at load carrying. In this paper the Method of Continuous Source Functions (MCSF) and Trefftz Radial Basis Functions (TRBF) will be presented. They are boundary meshless methods which do not need any mesh. The TRBF are source functions having their source points outside the domain. Special attention will be given to the application of the TRBF in the form of dipoles to the simulation of composites reinforced by fibres of finite length with large aspect ratio. In linear problems, only nodes on the domain boundaries and a set of source functions in points outside the domain are necessary to satisfy boundary conditions. Finally we will show MCSF to modelling of reinforced composites with glass fiber and epoxy matrix. For the sake of simplicity we consider only a patch of non-overlaying rows of fibers.

NOMENCLATURE

s	source point
f	field point
r	distance between s and t
$E_{pij}^{(D)}$	strain field
G	shear modulus
$S_{pij}^{(D)}$	stress field
$U_{pi}^{(F)}$	displacement field
$U_{pi,j}^{(F)}$	gradients of the displacement field
Greek symbols	
ν	Poisson's ratio

INTRODUCTION

Composite materials are now common engineering materials used in a wide range of applications. They play an important role in the aviation, aerospace and automotive industry, and are also used in the construction of ships, submarines, nuclear and chemical facilities, etc.

Composites are one of the biggest material inventions of the 20-th century. Composites with the highest stiffness and strength have continuous (glass, carbon, Kevlar or aramid) fibers embedded in thermoset (polyester or epoxy) matrix. Fibers carry mechanical load, matrix transmits load to the fibers, insures ductility and toughness, and protect the fibers against damage caused by manipulation or surroundings. The material of matrix defines the working temperature and manufacturing conditions.

Composite material is a material composed of two or more distinct phases (matrix phase and dispersed phase) and having bulk properties significantly different from those of any of the constituents. Composites can be considered as homogeneous at the macro-scale, but they are heterogeneous at micro-scale, i.e. at the scale comparable with geometrical sizes of the constituents. Properties of composites depend on the properties of the constituent phases, their geometry and relative amounts.

Unidirectional composites reinforced by long fibers are one kind of multicomponent composites. They can be considered as orthotropic and it is not necessary to perform homogenization in structural analysis. If multidirectional fiber-reinforced composites are analyzed, it is necessary to perform homogenization in order to calculate the components of the constitutive matrix, or tensor [3,4], which are required for further analysis of composite structures, for example damage and failure of composite structures.

GFRP composites are the cheapest and the most widely used [5]. One of the latest innovations is to use thermoplastics as a material for matrix production in the form of woven fabric from low-priced polypropylene and glass fibers, which can be thermoformed (melted polypropylene) or expensive high-temperature thermoplastic, e.g. PEEK, which provides better high temperature resistance and toughness to the composite. Continuous fibers are used for higher loads in GFRP composites.

Short glass fibers are cheaper and are used in higher volumes. GFRP composites are used in printed circuit boards, foremasts, bodies and interior panels of cars, household appliances, furniture and armatures.

Properties of fiber composites reinforced by long fibers significantly depend on the selection of fiber and matrix and way of how they are combined, fiber volume fraction, fiber length, fiber orientation, laminate thickness and presence of bond medium for improvement of fiber-matrix bond. Glass fibers have high strength at low costs; carbon fibers have very high strength, stiffness and low density. Kevlar fibers have high strength and low density; they prevent spreading of fire and they are penetrable by radio waves. Polyesters are the most often used matrices, because they offer good properties at relatively low costs. The best properties of epoxies and application of polyamides at temperatures predestinate them to special using, but they are expensive. Strength of composites increases by higher fiber volume fraction and fiber orientation parallel to load direction. By increasing length of fibers, the reinforcement is more effective at load carrying.

Shorter fibers are better for manufacturing and less expensive. The increase of laminate thickness leads to the reduction of composite strength and strength modulus, because of presence of defects. Influence of surroundings, like fatigue loading, humidity and temperature, decrease allowed strength.

Fiber-reinforced composites cover wide range of composite materials and structures such as laminated panels of unidirectional plies [6], sandwich structures with fiber-reinforced composite skins and flexible core, and fiber-metal laminates (FML). Although glass and carbon fibers is relatively common, polymer, metal and ceramic fibers are being used as reinforcement phase in specific applications. Fiber-reinforced polymer (FRP) matrix composites offer an alternate replacement material to high-strength metals.

Glass fibers fall into two categories: low-cost general-purpose fibers and premium special-purpose fibers. Over 90 % of all glass fibers are general- purpose products. These fibers are known by the designation E-glass. The remaining glass fibers are premium special-purpose products. Special-purpose fibers, which are of commercial significance in the market today, include glass fibers with high corrosion resistance (ECR-glass), high strength (S-, R-, and T E-glass), with low dielectric constants (D-glass), high-strength fibers, and pure silica or quartz fibers, which can be used at ultrahigh temperatures. These fibers are discussed in [8]. Table 1 provides properties of commercial glass fibers. The present work will use the method of continuous source functions (MCSF) and will show how the Trefftz Radial Basis Functions (TRBF), i.e. RBF, satisfying the governing equations (which can be the fundamental solutions, or more general functions, dipoles, dislocations, etc.) can be used to increase the efficiency of simulations composites reinforced by fibers [10,11]. The most efficient methods will be those which will approximate both domain variables and boundary conditions best. The TRBF are source functions having their source points outside the domain. Special attention will be given to the application of the TRBF in the form of dipoles to the simulation of composites reinforced by fibers with large aspect ratio (AR). Compared to the Method of Fundamental Solutions MFS [12, 13] which does not require any integration, the MCSF requires integration along 1D and 2D element. The integrals are quasi-singular and quasi-hyper-singular and the numerical integration is computationally cumbersome and inefficient. The analytic integration using symbolic manipulation is a very efficient tool used in the models. A relatively small number (usually fewer than 10) of elements in a fiber is necessary to obtain a good accuracy also with a large AR (e.g. 1: 100). The proposed method is not fully mesh less for these particular models as it requires 1D elements outside the 3D matrix domain, However, the model presents a significant reduction (even by several orders) of the resultant system of equations comparing to FEM, BEM, or other known mesh less methods and can be qualified as a Mesh Reducing Method (MRM).

Table 1. Properties of glass fibers

	Glass E Saint-Gobain Vertex. a.s.	Glass E literature data	Glass S	Glass C	Glass ACT	Silica glass
Diameter[μm]	5-16	9-13	9-13	9-13		8.9
Density [g/cm^3]	2.54	2.54	2.49	2.49	2.7	2.19
Elastic modulus [GPa]	73	72.4	85.5	69	75	69
Tensile strength [GPa]	2-4	up to 3.45	up to 4.6	up to 3	1.7	3.45
Extension [%]	1.8-3.2	4.8	5.7	4.8		5
Coefficient of thermal expansion	4.9	5	5.6	7.2		
Thermal conductivity	1	1	1	1		
Softening point [°C]	800	800	970	750		980
Relative permittivity at frequency 1 MHz	5.9-6.4		5-5.4			3.78
Dissipation factor (tan δ) at 10 GHz		0.0039				0.002

NUMERICAL MICROMECHANICAL MODELS

Direct numerical analysis of the material behavior, which takes into account both macroscopic boundary conditions and micro-geometries, would require very large computational resources. To overcome this problem, several strategies are employed:

- Multi-scale modeling: the material behavior is modeled at the scale levels of both microstructures and the sample; the lower scale model is subjected to the boundary conditions, acting on the macro-model, while the mechanical behavior of the material in the macro-model is determined taking into account finer micro-scale features incorporated into the micro-model. In the simplest models, upper scale–lower scale relationships are one-sided, while more sophisticated models allow simultaneous global-local analysis at several levels.
- Homogenization and averaging of properties and micro-fields: the material is considered as a homogeneous equivalent medium at the macro-level, and the effective properties of the medium are determined on the basis of the analysis of the microstructure, micro-geometry and properties of the materials.

Some analytical and numerical techniques have been used for prediction and characterization of composite microstructure behavior. Analytical methods provide reasonable prediction for relatively simple configurations of the phases. Complicated geometries, loading conditions and material properties often do not yield analytical solutions, due to complexity and number of equations. In this case, numerical methods are used for approximated solutions, but they still make some simplifying assumptions about the microstructures of heterogeneous multiphase materials. One of such methods is the finite element method (FEM) [18]. Computational simulations can help in the understanding, analyzing and designing of fiber-reinforced composite materials (FRCM). Among the existing of FEM, boundary element method (BEM) [19] and various mesh less methods [20, 21] can be used for simulating behavior of the FRCM. However, these classical numerical methods are suitable for the lowest scale simulation only. It is worth noting that in FRCM with short fibers the fiber aspect ratio (length to diameter) of short fibers is often 100, 1000, or even larger, which results in very large gradients in strain and stress fields along fibers and in matrix. FEM is an alternative approach to solving the governing equations of a structural problem. In the FEM the equations are formulated for each finite element and assembled in order to obtain a solution for the whole body instead of solving a uniform mathematical problem for the entire body. This method involves modeling the structure, using interconnected elements called finite elements, consisting of interconnected nodes and/or boundary lines and/or surfaces that are directly or indirectly linked with other elements via interfaces. Each finite element has an assumed displacement field. FEM requires selection of appropriate elements of suitable size and distribution (the FEM mesh).

Displacement function and material properties are associated with each finite element. Boundary conditions and loading define behavior of each node and these are expressed in matrix notations. FEM combines a model in the form of microstructures with fundamental material properties such as elastic modulus or coefficient of thermal expansion of constitutive phases as a basis for understanding material behavior. Solutions are stress and strain data for each node in the system and they are summarized according to usual criteria. It is well known that using FEM hundreds of elements are necessary to achieve required accuracy even for simple problem [22] where 50 000 to 100 000 trilinear elements were used for problem containing one spherical particle in the matrix. All the methods mentioned above require millions or even billions of equations after numerical discretization to obtain a sufficiently accurate solution of this kind of computer simulation problem.

Method of Continuous Source Functions (MCSF)

A composite material with micro/nano structure with regularly distributed reinforcing fibers of unique dimension is to be modeled. Due to very small dimension of particles, stress and strain field can be considered to be homogeneous in the whole domain and all stresses and strains can be imagined as a part on the constant part, which introduces the state of the matrix without stiffening and the local part, due to stiffening effect by the fibers.

Let's assume that cross sectional dimensions of fiber are much smaller than its length, tensional stiffness of fiber is much higher than stiffness of matrix, fiber is straight and ideal cohesion forces are assumed in the present model. Then the action of the fiber can be introduced by zero strains in longitudinal direction of fiber boundary and zero difference of displacements in directions perpendicular to the fiber axis. AR (length to radius) of fiber is considered to be very large (100:1, 1000:1 in examples used here).

Because of large aspect ratio, continuity of strains between matrix and fiber along the fiber axis can be simulated by continuously distributed source functions (forces, dipoles, dislocations, etc.) [23, 24] as they are known from the potential theory (Fig.1). Continuous source functions enable to simulate the continuity conditions with much reduced collocation points along the fiber boundary.

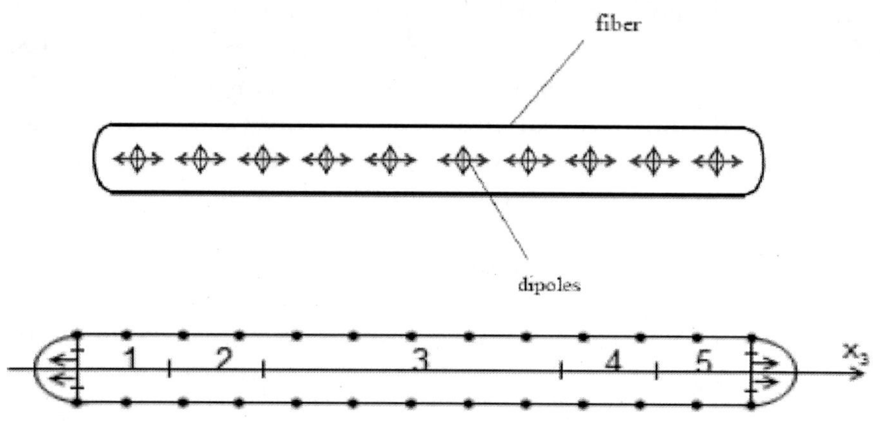

Figure 1. Continuous force/dipole model for fiber reinforcing element

All displacement, strain and stress fields will be split into homogeneous part corresponding to constant stress-strain in matrix without the reinforcement. For simplicity an isotropic material properties are assumed in this paper.

Field of displacements in elastic continuum caused by unit force acting in direction of the axis x_p is given by Kelvin solution

$$U_{pi}^{(F)} = -\frac{1}{16\pi G(1-\nu)}\frac{1}{r}\left[(3-4\nu)\delta_{ip} + r_{,i}\, r_{,p}\right] \tag{1}$$

where i denotes the x_i coordinate of displacement, G and v are shear modulus and Poisson's ratio of material of the matrix (isotropic material is considered here), r is the distance between the source point s, where the force is acting with the field point t, where the displacement is introduced, i.e.

$$r = \sqrt{r_i r_i}\ ,\ r_i = x_i(t) - x_i(s) \tag{2}$$

The summation convection over repeated indices acts and

$$r_{,i} = \partial r / \partial x_i(t) = r_i / r \tag{3}$$

is the directional derivative of the radius vector r. Gradients of displacement fields are corresponding derivatives of the field in Eq. (1) in the point t

$$U_{pi,j}^{(F)} = -\frac{1}{16\pi G(1-v)r^3}\left[(3-4v)\delta_{pi}r_{,j} - \delta_{pj}r_{,i} - \delta_{pj}r_{,p} + 3r_{,i}r_{,j}r_{,p}\right] \tag{4}$$

The second derivative of the radius vector of n-th power is

$$\left(r_{,k}^{n}\right)_{,j} = \frac{n}{r}\left(r_{,k}^{n-1}\delta_{jk} - r_{,j}r_{,k}^{n}\right) \tag{5}$$

Strains are

$$E_{pij}^{(F)} = \frac{1}{2}\left(U_{pi,j}^{(F)} + U_{pj,i}^{(F)}\right) = -\frac{1}{16\pi G(1-v)}\frac{1}{r^2}\left[(1-2v)(\delta_{pi}r_{,j} + \delta_{pj}r_{,i}) - \delta_{ij}r_{,p} + 3r_{,i}r_{,j}r_{,p}\right] \tag{6}$$

and the ij stress components of this field are

$$S_{pij}^{(F)} = 2GE_{pij}^{(F)} + \frac{2G\nu}{1-2\nu}\delta_{ij}E_{pkk}^{(F)} = \frac{1}{8\pi(1-\nu)}\frac{1}{r^2}\left[(1-2\nu)(\delta_{ij}r_{,p} - \delta_{jp}r_{,i} - \delta_{ip}r_{,j}) - 3r_{,i}r_{,j}r_{,p}\right] \tag{7}$$

where δ_{ij} is the Kronecker delta.

The displacement field of dipole can be obtained from the displacement field caused by the force by differentiating it in the direction of the acting force, i.e.

$$U_{pi}^{(D)} = U_{pi,p}^{(F)} = -\frac{1}{16\pi G(1-\nu)}\frac{1}{r^2}\left[3r_{,i}r_{,p}^2 - r_{,i} + 2(1-\nu)r_{,p}\delta_{ip}\right] \tag{8}$$

The summation convection does not act over the repeated indices p here and in the following relations, too. Gradients of the displacement field are

$$U_{pi,j}^{(D)} = -\frac{1}{16\pi G(1-\nu)}\frac{1}{r^3}\left[-15r_{,i}r_{,j}r_{,p}^2 + 3r_{,i}r_{,j} + 2(1-2\nu)\delta_{ip}(\delta_{jp} - 3r_{,j}r_{,p}) + \right.$$

$$\left. + 6r_{,i}r_{,p}\delta_{jp} + \delta_{ip}(3r_{,p}^2 - 1)\right] \tag{9}$$

and corresponding strain and stress fields are

$$E_{pij}^{(D)} = \frac{1}{2}\left(U_{pi,j}^{(D)} + U_{pj,i}^{(D)}\right) = -\frac{1}{16\pi G(1-\nu)}\frac{1}{r^3}\left[-15r_{,i}r_{,j}r_{,p}^2 + 3r_{,i}r_{,j} + \right.$$

$$\left. + 2(1-2\nu)\delta_{ip}\delta_{jp} + 6\nu(\delta_{ip}r_{,j}r_{,p} + \delta_{jp}r_{,i}r_{,p}) + \delta_{ij}(3r_{,p}^2 - 1)\right] \tag{10}$$

$$S_{pij}^{(D)} = 2GE_{pij}^{(D)} + \frac{2G\nu}{1-2\nu}\delta_{ij}E_{pkk}^{(D)} = -\frac{1}{8\pi(1-\nu)}\frac{1}{r^3}\left[(1-2\nu)(2\delta_{ip}\delta_{jp} + 3r_{,p}^2\delta_{ij} - \delta_{ij}) + \right.$$

$$\left. + 6\nu r_{,p}(r_{,i}\delta_{jp} + r_{,j}\delta_{ip}) + 3(1-5r_{,p}^2)r_{,i}r_{,j}\right] \tag{11}$$

The displacements, Eq. (1), caused by force are weak singular, the displacement gradients, strains and stresses are strong singular. Fields defined by dipole have one order higher singularity (strong singularity in the displacement field and hyper singularities in the strain and stress fields). The derivatives in perpendicular direction to the force define the force couple [25]. The displacement field for the couple is

$$U_{pi}^{(C)} = U_{pi,i}^{(F)} = -\frac{1}{16\pi G(1-v)}\frac{1}{r^2}\left[3r_{,p}r_{,i}^2 - r_{,p} + 2(1-v)r_{,i}\delta_{ip}\right] \tag{12}$$

and corresponding strain and stress fields are

$$E_{pij}^{(C)} = \frac{1}{2}\left(U_{pi,j}^{(C)} + U_{pj,i}^{(C)}\right) =$$
$$-\frac{1}{16\pi G(1-v)}\frac{1}{r^3}\left[-15r_{,p}r_{,j}r_{,i}^2 + 3r_{,p}r_{,j} + 2(1-2v)\delta_{ip}\delta_{ij} + 6v(\delta_{ip}r_{,i}r_{,j} + \delta_{ij}r_{,i}r_{,p}) + \delta_{jp}(3r_{,i}^2 - 1)\right] \tag{13}$$

$$S_{pij}^{(C)} = 2GE_{pij}^{(C)} + \frac{2Gv}{1-2v}\delta_{ij}E_{pkk}^{(C)} = -\frac{1}{8\pi(1-v)}\frac{1}{r^3}\Big[(1-2v)(2\delta_{ip}\delta_{ij} + 3r_{,i}^2\delta_{jp} - \delta_{jp}) +$$
$$+ 6vr_{,i}(r_{,j}\delta_{ip} + r_{,p}\delta_{ij}) + 3(1-5r_{,i}^2)r_{,p}r_{,j}\Big] \tag{14}$$

The RBF can be also used for simulation of the interaction between matrix and particles with very large aspect ratio such as composites reinforced by short fibers where the aspect ratio can be 1000:1 or even larger. The inter-domain boundary conditions can be simulated by 1D distribution of TRBF (source functions) along the fiber axis. When TRBF are approximated by polynomials then the problems lead to evaluation of the following integrals

$$\int_a^b \frac{x_s^n (x_s - x_f)^p}{\left(y^2 + x_s - x_f^2\right)^{\frac{m}{2}+r}} \mathrm{d}x_s = f(x_f) \tag{15}$$

where x is the coordinate along the fiber axis, the subscripts s and f denote the source and the field point and exponents are integer numbers and y is the distance of the field point from the source point. For computational purpose the integral in Eq. (15) is transformed to

$$\int_{a+x_f}^{b+x_f} \frac{(x+x_f)^n x^p}{(y^2+x^2)^{\frac{m}{2}+r}} \mathrm{d}x = f(x_f) \qquad (16)$$

Numerical integration of such integrals would be computationally very laborious because of quasi-singularities and quasi-hyper-singularities in the integrals. However, analytical calculation of integrals containing the kernel function and polynomial approximation of unknown function is very elegant way of numerical calculation of the integrals, if the axis of fiber is straight, i.e. the value of y is constant in the integrals above.

Fiber-matrix Interaction and Failure Criteria

When a particle reinforced composite is subjected to mechanical load, the entire load is carried initially by the matrix.

The matrix is deformed, and its elastic deformation and plastic flow lead to load transfer to the particles. The extreme shear forces between fiber and matrix can lead to de-bonding or to de-cohesion and in the end parts and also in the middle of the fiber close to another fiber in materials reinforced with fibers or nanotubes, which are typical and very efficient novel reinforcing materials. This middle part of the fiber will carry the largest forces, which can lead to its fracture. The main mechanisms of the influence of the particles on the deformation of particle reinforced composites are load transfer (i.e. the matrix transfers some of the applied stress to the particles, which carry a part of the load), and constraining the matrix deformation by the particles [26].

The basic assumptions in unidirectional lamina owing to the tensile and compressive loadings are [27]:

- Fibers are uniformly distributed throughout the matrix.
- Perfect bonding exists between fibers and matrix.

- Matrix is free of voids.
- Applied force is either parallel to or normal to the fiber direction.
- Lamina is initially in stress-free state (i.e., no residual stresses are present in the fibers and in the matrix).
- Both fibers and matrix behave as linearly elastic materials.

Tensile load applied to discontinuous fibrous lamina is transferred to the fibers by a shearing mechanism between fibers and matrix. Micromechanical models are based on the microstructure and mechanical characteristics of the constituents (fibers and matrix) of the composite. These models are used to predict effective mechanical properties of the composite and provide the insight into the contributions of each constituent to properties and interactions between fibers and matrix.

However, in mechanical analysis of composite structures, microscopic models may require expensive computations and are hampered by insufficient data of the constituents [5]. Also, for such analysis, it is important to insure that the model employed can adequately describe the macroscopic behavior of the composite. Therefore, macroscopic models of the composite are generally considered to be more suitable and preferable than microscopic models for structural analysis.

For unidirectional discontinuous fibers tensile load is transferred to the fibers by shearing mechanism between fibers and matrix. Since the matrix has lower modulus, the longitudinal strain in the matrix is higher than that in adjacent fibers. If a perfect bond is assumed between two constituents, the difference in longitudinal strains creates a shear stress distribution across the fiber–matrix interface. Ignoring the stress transfer at the fiber end cross sections and the interaction between the neighboring fibers, we can calculate the normal stress distribution in discontinuous fiber by a simple force equilibrium analysis [25]. Then the maximum fiber stress that can be achieved at a given load is

$$\left(\sigma_f\right)_{max} = 2\tau_i \frac{L_t}{D_f} \tag{17}$$

From (17) for a given fiber diameter and fiber-matrix interfacial condition, the critical fiber length L_c is calculated

$$L_C = \frac{\sigma_{fu}}{2\tau_i} D_f \qquad (18)$$

Where

σ_{fu}=ultimate tensile strength of the fiber

L_t=maximum fiber length in which the maximum fiber stress is achieved

τ_i= shear strength of the fiber-matrix interface or shear strength of the matrix adjacent to the interface

D_f=fiber diameter

Note that for determining the fiber-matrix interfacial shear strength τ_i a single fiber fragmentation test is used, which is based on the observation that fibers will not break if their length is less than the critical value. In our case we use results from numerical simulation by MCSF for determining τ_i.

DIRECT APPLICATION OF MCSF TO DESIGN OF GFRP COMPOSITE

In this example, we show how the above results can be used for designing GFRP composite with short fibers. For the sake of simplicity we consider only a patch of non-overlaying rows of fibers. Material properties of GFRP composite are listed in Table 2. In this example the maximum elastic modulus of the epoxy matrix was selected as E= 4000 MPa and Poisson's ratio $\nu = 0.35$. The matrix was reinforced by a patch of straight rigid cylindrical fibers. The length of fibers was L= 1 mm and the radius R= 0.009 mm. The distance between fibers was $\Delta_1=\Delta_2=\Delta_3$=0,144 mm. The patches of fibers consisted of 10 x 10 x 5 fibers in the presented examples. The domain is supposed to be loaded by far field stress $\sigma_{33\infty}$ =10 MPa in the (x_3)-direction.

Table 2. Material properties of composite

	Density □ (g/cm^3)	Young's modulus *E*(GPa)	Maximum specific module *E*/□ (MNm/kg)	Ultimate Strength σ_u (MPa)	Maximum Specific strength σ_u/□(kNm/kg)	Diameter d (μm)
E-glass	2.54	72.4	29	3450	1646	5-18
Epoxy matrix	1.1-1.6	3-6	-	30-100	-	-

The strains along the fiber and in plane perpendicular to the fiber are plotted in Fig. 3 and Fig. 4. From these figures we can see that maximum value of strain along the fiber is approximately $□_{33}$= -0.125 and maximum absolute value of strain in plane perpendicular to the fiber is approximately $□_{11}$ =$□_{22}$= -7.0e-03. The stresses along the fiber and in plane perpendicular to the fiber are plotted in Fig. 5 and Fig. 6. From these figures we can see that maximum value of stress along the fiber is approximately σ_{11}=-350 MPa and in plane perpendicular to the fiber σ_{22} = -19 MPa. The shear stresses along the fiber and in plane perpendicular to the fiber are plotted in Fig. 7 and Fig. 8. From these figures we can see that maximum value of shear stress along the fiber is approximately τ_{31}= τ_{23}= -400 MPa and in plane perpendicular to the fiber is approximately τ_{12}=-90 MPa.

From shear stress along the fiber and using equation (17) we can calculate the critical fiber length L_c

$$L_c = \frac{\sigma_{fu}}{2\tau_i} D_f = \frac{3450}{2.400} 0.018 = 0.07625 \text{ mm}$$

where $\tau_i = |\tau_{31}| = |\tau_{23}|$ =400 MPa is taken from simulation (fig.7) and σ_{fu} is taken from table 2. From the result we can see that L_f □ L_c , the maximum fiber stress may reach the ultimate fiber strength over much of its length. However, over a distance equal to $L_c/2$ from each end, the fiber remains less effective. We note on the basis of [25] that for effective fiber reinforcement, i.e. for using the fiber due to its potential strength, one must select L_f □ □ L_c. This is fulfilled in our case.

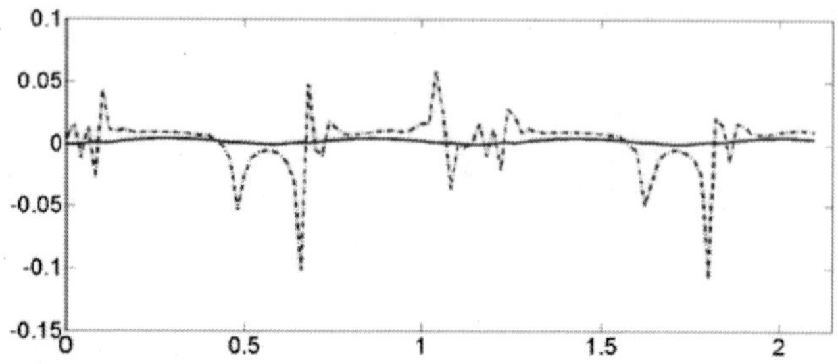

Figure 3. Strain along/parallel to the fiber

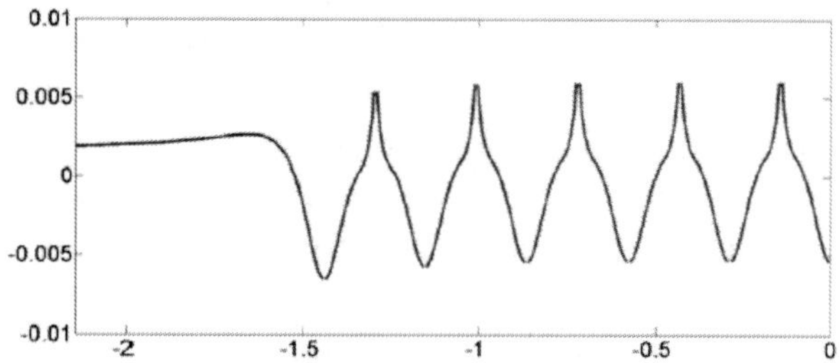

Figure 4. Strain in plane perpendicular to the fiber

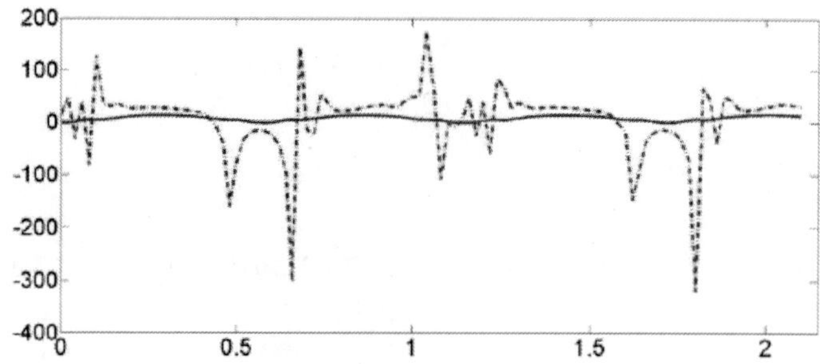

Figure 5. Stress along/parallel to the fiber

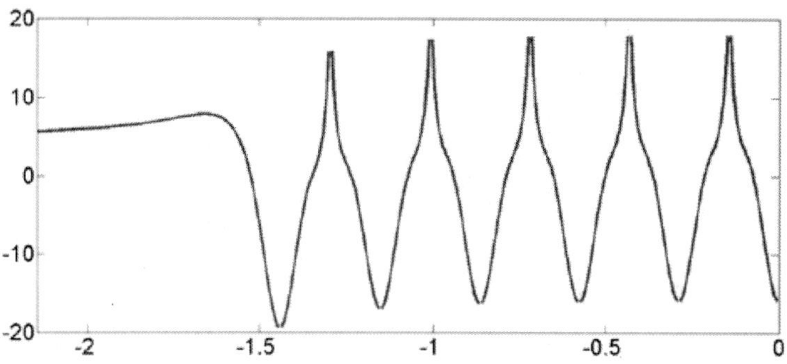

Figure 6. Stress in plane perpendicular to the fiber

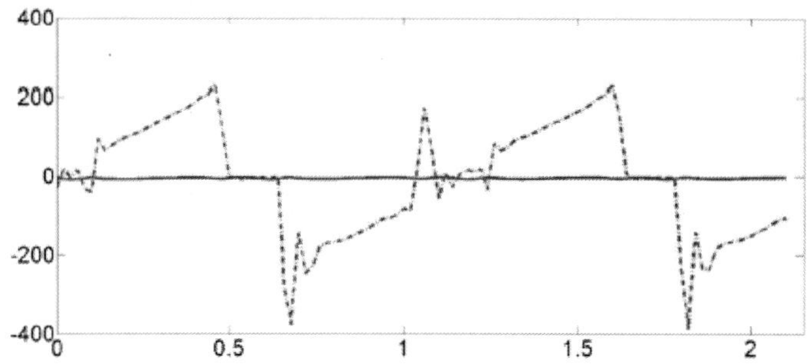

Figure 7. Shear stress along/parallel to the fiber

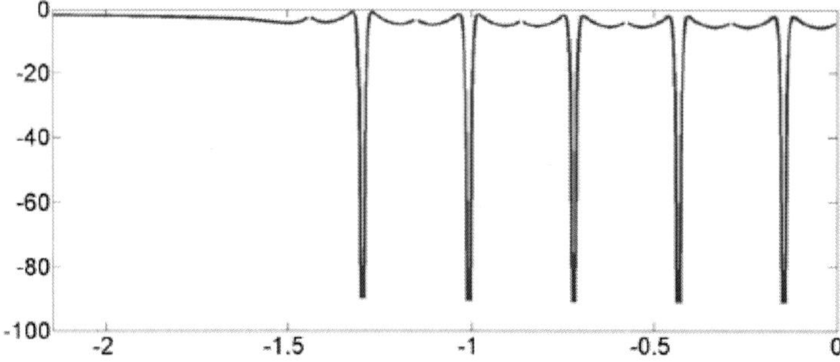

Figure 8. Shear stress in plane perpendicular to fiber

CONCLUSION

In the first part of this chapter the formulations based on the MCFS and the RBF for analysis of an elastic medium containing stiff reinforcing fibers are applied. The MCSF enables to simulate both the interaction of matrix with these fibers and the fiber with other fibers very effectively. In the example, we showed how MCFS can be used for designing GFRP composite with short fibers. For the sake of simplicity we considered only a patch of non-overlaying rows of fibers.

From the results we can see that fiber length is much greater than the critical length, so the maximum fiber stress may reach the ultimate fiber strength over much of its length. Optimal aspect ratio of fibers can be found for specific working stress/strain conditions in each part of structure. This is another attractive property of this kind of composite materials. Main advantage of the MCFS is the fact that numerical models used for simulation of all matrix-fiber and fiber-fiber interactions have to keep the gradients in order to correctly estimate the interactions. FEM can smooth out the gradients and so very fine meshes would be necessary for numerical models.

ACKNOWLEDGEMENTS

The authors gratefully acknowledge the support by the Slovak Grant Agency VEGA 1/1226/12 and Slovak Science and Technology Assistance Agency registered under number APVV-0169-07.

REFERENCES

1. Voyiadjis, G. Z., Kattan, P. I., 2009. A Comparative Study of Damage Variables in Continuum Damage Mechanics, International Journal of Damage mechanics 18, p. 315 340.
2. Kaw, A. K., 2006. Mechanics of Composite Materials. 2nd ed. ,CRC Press, Boca Raton.
3. Mishnaevsky, L., Jr. 2007. Computational Mesomechanics of Composites: Numerical analysis of the effect of microstructures of composites on their strength and damage resistance, John Wiley & Sons. Ltd, Chichester, U. K.
4. Kormaníková, E., Kostrasová, K., 2011. Analysis of laminates and sandwiches, in "Selected Chapters of Mechanics of Composite Materials I",

E. Kormaníková, et al., Editor, Technical University of Košice, p 265.

5. Siáková, A., 2011. Introduction to composite materials, in "Selected Chapters of Mechanics of Composite Materials I", E. Kormaníková, et al., Editor, Technical University of Košice, p. 265.
6. Wallenberger, F. T., Watson, J. C., Li, H. Glass Fibers, ASM Handbook, ASM International 21.
7. Huang, X., 2009. Fabrication and properties of carbon fibers, Materials, p. 2369-2403.
8. Murinková, Z., Kompiš, V., Štiavnický, M., 2008. Trefftz functions for 3D stress concentration problems, Computer Assisted Mechanics and Engineering Sciences 15, p. 305-318.
9. Kompiš, V., Mur□inková, Z., Rjasanow, S., Grzibovskis, R., Qin, Q.H. 2010. Computational simulation methods for fiber reinforced composites, Frontiers of Architecture and Civil Engineering in China 4, p. 396-401.
10. Golberg, M. A., Chen, C. S., 1998. The Method of Fundamental Solutions for Potential, Helmholtz and Diffusion Problems, Boundary Integral Methods - Numerical and Mathematical Aspect 1, p. 103-176.
11. Karageorghis, A., Fairweather, G., 1989. The method of fundamental solutions for the solution of nonlinear plane potential problems, IMA Journal Numerical Analysis 9, p. 231-242.
12. Laš, V, Zem□ík, R., 2008. Progressive Damage of Unidirectional Composite Panels, Journal of Composite Materials 42, p. 25-44.
13. Tay, T. E., Liu, G., Yudhanto, A., Tan, V.B.C.A, 2008. Micro Macro Approach to Modeling Progressive Damage in Composite Structures, International Journal of Damage Mechanics 17, p. 5-28.
14. Kormaníková, E., Riecky, D., Žmindák, M., 2011. Strength of composites with fibers, in "Computational Modelling and Advanced Simulations", J. Murín, V. Kompiš, V. Kutiš, Editors. Springer Science + Business Media B.V., p. 364.
15. Puck, A., Schurmann, H., 2002. Failure analysis of FRP laminates by means of physically based phenomenological models, Composite Science and Technology 62, pp. 1633-1662.
16. Bathe, K. J., 1996, Finite Element Procedures, Prentice Hall, New Yersey.
17. Ma, H., Qin, Q. H., 2007. Boundary point method for linear elasticity based on direct formulations of conventional and hypersingular boundary integral equations, Engineering Analysis with Boundary Elements 33, pp. 410-419.
18. Chen, Y., Lee, J. D., Eskandarian, A., 2006. Meshless Methods in Solid Mechanics, Springer, Berlin.
19. Sladek, J., Sladek, V., Mang, H.A., 2002. Meshless formulations for simply supported and clamped plate problems, International Journal for Numerical Methods in Engineering 55, pp. 359-375.
20. Filip, C., Garnier, B., Danes, F., 2005." Prediction of the effective thermal conductivity of composites with spherical particles of higher thermal conductivity than the one of the polymer matrix.," Proceedings of the

COMSOL Multiphysics User's Conference.

21. Blokh, V. I., 1964. Theory of Elasticity, University Press, Kharkov.
22. Kachanov, M., Shafiro, B., Tsukrov, I., 2003. Handbook of Elasticity Solutions, Kluver Academic Publishers, Dordrecht.
23. Rjasanow, S., 2007. "Fast and accurate boundary element method," APCOM›07-EPMESC XI.
24. Luo, H. A., Chen, Y., 1991. Matrix Cracking in Fiber-Reinforced Composite Materials, Journal of Applied Mechanics 58, pp. 846-848.
25. Mallick, P.K., 2007. Fiber-reinforced composites: materials, manufacturing and design, CRC Press, Boca Raton.
26. Barbero, E.J., 2008. Finite Element Analysis of Composite Materials, CRC Press, Boca Raton.2008.
27. Barbero, E.J., de Vivo, L., 2001. A Constitutive Model for Elastic Damage in Fiber-Reinforced PMC Laminae, International Journal of Damage Mechanics 10, pp. 73-93.
28. Jirásek, M., Zeman, J., 2006. Deformation and damage of materials (in Czech), Czech Technical University, Prague.
29. Barbero, E.J., Lonetti, P., 2002. An inelastic damage model for fiber reinforced laminates, Journal of Composite Materials 36, pp. 941-962.
30. Boresi, A.P., Schmidt, R.J., 2003. Advanced mechanics of materials, John Wiley & Sons. Ltd, Chichester, U. K.
31. Zhang, Y. X, Chang, C. H., 2009. Recent developments in finite element analysis for laminated composite plates, Composite Structures 88, pp. 147-157.
32. Voyiadjis, G.Z., Kattan, P.I., 2009. A Comparative Study of Damage Variables in Continuum Damage Mechanics, International Journal of Damage Mechanics 18, pp. 315-340.
33. Guiamatsia, I., Falzon, B.G., Davies, G.A.O., Iannucci, L., 2009. Element-free Galerkin modelling of composite damage, Composites Science and Technology 69, pp. 2640-2648.

Chapter 4

NDT OF FIBER-REINFORCED COMPOSITES WITH A NEW FIBER-OPTIC PUMP–PROBE LASER-ULTRASOUND SYSTEM

Ivan Pelivanov [a,b], Takashi Buma[c], Jinjun Xia[a], Chen-Wei Wei[a], Matthew O'Donnell[a]

[a]Department of Bioengineering, University of Washington, Seattle, WA, USA
[b]International Laser Center, Moscow State University, Moscow, Russian Federation
[c]Union College, Schenectady, NY, USA

ABSTRACT

Laser-ultrasonics is an attractive and powerful tool for the non-destructive testing and evaluation (NDT&E) of composite materials. Current systems for non-contact detection of ultrasound have relatively low sensitivity compared to contact peizotransducers. They are also expensive, difficult to adjust, and strongly influenced by environmental noise. Moreover, laser-ultrasound (LU) systems typically launch only about 50 firings per second, much slower than the kHz level pulse repetition rate of conventional systems.

As demonstrated here, most of these drawbacks can be eliminated by combining a new generation of compact, inexpensive, high repetition rate nanosecond fiber lasers with new developments in fiber telecommunication optics and an optimally designed balanced probe beam detector. In particular, a modified fiber-optic balanced Sagnac interferometer is presented as part of a LU pump–probe system for NDT&E of aircraft composites. The performance of the all-optical system is demonstrated for a number of composite samples with different types and locations of inclusions.

INTRODUCTION

For more than thirty years, optical generation and detection of ultrasound (US) has been used for NDT&E of composite materials. All-optical approaches can produce non-contact systems for US inspection, a clear advantage over more traditional contact methods based on piezoelectric transducers. The history of developments in this area is well described in a series of review articles [1], [2], [3] and [4], with particular note of the seminal work by J.-P. Monchalin's group [5], [6], [7] and [8], and some studies on adapting Sagnac interferometry to US detection [9], [10], [11], [12], [13] and [14]. Alternate optical detectors with sensitivity rivaling the best piezoelectric transducers have also recently been reported by Guo [15],[16] and [17] and Razansky [18], but these schemes still require immersion. To summarize, there are currently no non-contact techniques for US detection approaching the sensitivity of the best contact detectors. Thus, there is a real opportunity to advance the state of the art in non-contact ultrasound systems by developing low-cost, non-contact optical detectors with sensitivity approaching the best piezoelectric transducers.

Our approach to non-contact LU systems leverages recent progress in fiber and diode laser systems and telecommunication optics, and addresses a number of the limitations of LU systems commonly used for NDT&E of composite materials in the aerospace industry. In particular, our design goals are to:

- reduce the footprint of the detection system;
- decrease system cost and complexity;
- increase scan speeds;

- minimize the impact of environmental noise;
- provide detection sensitivity comparable to that of piezoelectric transducers.

To meet these goals, high-energy solid-crystal and CO_2 lasers cannot be used for optical generation of US because of their low repetition rate (usually less than 100 Hz), bulk, and cost. Instead, we use modern diode-pumped or fiber lasers with a typical pulse energy of ~1 mJ operating at repetition rates up to 100 kHz. As demonstrated below, they are quite capable of exciting strong acoustic transients by optical absorption at the composite surface; their high repetition rate can dramatically increase scan speed; and their size can dramatically reduce system footprint. In addition, these lasers are much more flexible and relatively inexpensive.

Most non-contact optical detection schemes rely on some form of interferometry. Recent advances in continuous wave (CW) fiber lasers (for highly coherent output) and super luminescent diodes (SLD) can be exploited for more sensitive detection based on optical interferometry. Both sources are low noise, compact, and quite inexpensive. Below we show that an SLD is nearly ideal when a relatively low coherence source is required for interferometry.

Most interferometers are highly sensitive to environmental noise. Thermal fluctuations from any source, such as room temperature shifts or changes in the temperature of the sample surface induced by pumping laser radiation and the laser pump itself, can change resonance conditions in the interferometer cavity. Similarly, vibrations from mechanical and acoustic sources can change resonance conditions. As the resonance moves in wavelength, or reference beam phase, the sensitivity of the detector can be significantly degraded unless an active, robust stabilization system is included to compensate. A number of stabilization algorithms have been developed over the last twenty years depending on the type of interferometer [1],[2] and [3] but all are relatively slow, cannot provide complete rejection of background noise, and significantly increase the size, complexity, and cost of the interferometer.

For LU inspection of real composite structures used in the aerospace industry, reflections from rough surfaces must be processed effectively. A specific confocal Fabry–Perot interferometer design significantly reduced the influence of the light collection system [5],[6],[7] and [8]. Other approaches aimed at minimizing the effects of speckle structure include modifications of a Sagnac interferometer [9],[10], [11], [12], [13] and [14] and interferometer schemes employing photorefractive crystals [6], [19],[20] and [21]. However, the detection sensitivity of all these optical techniques does not approach that of a well-designed, contact piezoelectric transducer. Indeed, there is still need for an LU system with overall sensitivity rivaling the best piezoelectric systems.

To handle rough surfaces and simultaneously minimize the effects of environmental noise, we employed a Sagnac-based interferometer [9], [10], [11], [12], [13] and [14] measuring vibration speed instead of displacement. This approach has no reference arm, i.e. both interfering beams reflect from the sample surface and, therefore, no stabilization is required. In addition, the differential Sagnac approach is relatively insensitive to the roughness of the sample surface over the extent of the beam probing that surface. This is in stark contrast to an interferometer with a reference arm using a highly coherent source, such as Michelson or conventional Fabry–Perot schemes, in which any surface heterogeneities produce speckle structure in the reflected beam. Most schemes utilizing a reference beam are dramatically affected by speckle structure in the reflected beam and must use only the central speckle for stable operation. This problem can be partially eliminated by finely focusing the probe beam to the sample surface, but it does not help in all cases and creates another problem - fine probe beam focus adjustment. As shown below (see Section 2), our interferometric detector includes focusing of the probe beam onto the sample surface for better light collection from a rough surface. However, since the Sagnac approach does not have a reference arm, the reflected beam mode structure does not influence ultrasound signal characteristics and only changes the overall amplitude of the recorded signal.

Finally, to meet all of our design goals, the interferometric detector must approach the sensitivity of piezoelectric transducers. No matter what modality is used, the fundamental limit on sensitivity is determined by the Johnson–Nyquist noise associated with detection of an acoustic signal over a finite spatial aperture and finite signal bandwidth [2], [3], [12], [22] and [23]. As will be demonstrated below, the non-contact detector developed for our LU system approaches the Johnson-Nyquist noise limit and is as sensitive as the best acoustic detectors operating over the same detection area and signal bandwidth.

MATERIALS AND METHODS

Samples of Fiber Reinforced Composites

Samples used in this study were provided by Boeing Research & Technology and were typical fiber reinforced graphite-epoxy composites employed in aircraft at Boeing.

During system testing, we scanned a series of samples. Each one could be imaged with high SNR and good visibility of all significant defects. Without loss of generality, here we present results obtained on only three representative samples with artificial inclusions embedded in the structure. All samples (sound speed was measured to be very close to 3000 m/s, volume density is about 1600 kg/m^3) were 3.6 mm thick and contained 19 individual layers (i.e. plies). Artificial defects were made from about 20 μm thick brass foil, tape, and polymer foreign materials and were placed at different depths from the surface. Samples differed one from another by the depth of inclusion locations (close to and far from the nominal "front" surface of the sample, and in the middle, corresponding to samples UW-N, UW-F and UW-M). A photograph and diagram of inserted inclusions for sample UW-M are presented in Fig. 1a.

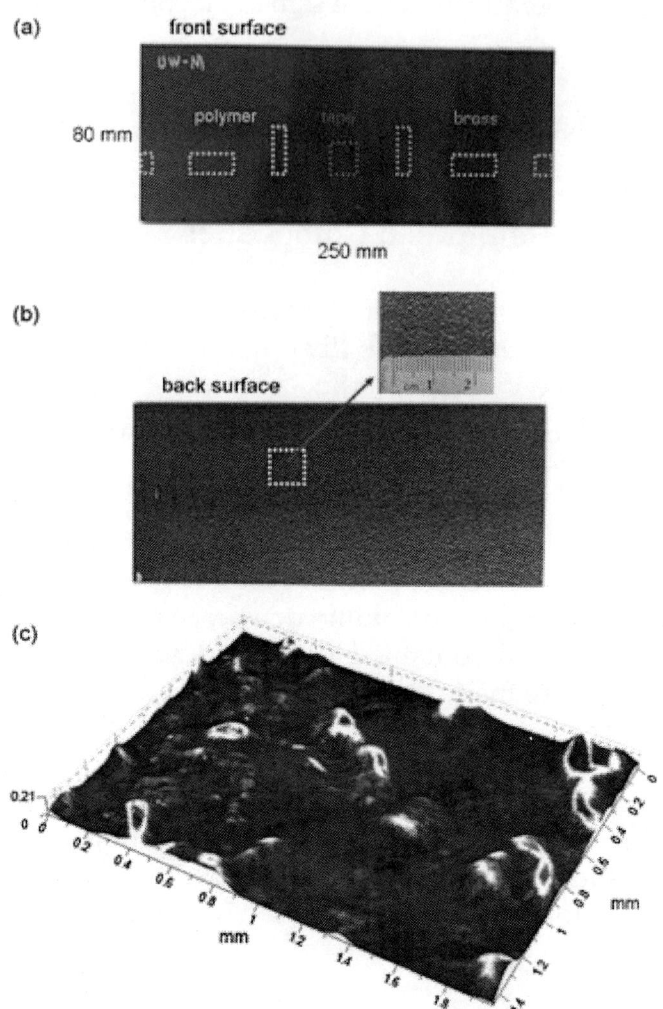

Figure-1 (a) Diagram of the composite sample (UW-M). Dashed rectangles in the upper picture represent the inclusions, their size and location. (b) Photo of the back surface (the enlarged photo indicates sample roughness in the region defined by the white rectangle). (c) Back wall roughness profile measured with a profilometer over a typical small region.

Composite samples were not transparent for both pump and probe light. Analyzing the amplitude and the profile of the generated OA signals, we estimate a light absorption coefficient to be about 200 cm^{-1} for 1064 nm wavelength.

As noted above, surface roughness is a critically important parameter affecting the sensitivity of interferometry. Clearly, the composite surface is far from mirror-like. The "front" surface was smoother. Based on conventional profilometry, the roughness of this surface (mean height of heterogeneities) is about 4 μm with many relatively flat regions. In contrast, the "back" surface is very rough, as seen in the photo of Fig. 1b and the profilometry image of a small, typical region of this surface in Fig. 1c. The mean height of the roughness profile for the back surface is more than 130 μm and reaches more than 200 μm at some points. Most experiments used the front as the source/detection surface, but some results will also be presented below using the back surface.

All Optical Inspection System

Based on the design goals presented in Section 1, the LU inspection system illustrated in Fig. 2a was constructed. Optical generation is driven by a compact, inexpensive fiber laser operating at quite low pulse energies (0.6 mJ) but at very high pulse repetition rates (variable up to 76 kHz) rather than a bulky, high cost, low repetition-rate CO_2 or solid crystal laser. Pulsed radiation at a 1064 nm wavelength is delivered to the sample from the optical head of the fiber laser (Model G3, 40 W pulsed fiber laser, spilasers.com). Outgoing collimated radiation is focused to the sample surface using one convex lens (150 mm focal distance) attached to the pump laser head that allows the beam diameter at the sample surface to be easily varied. The laser beam diameter is optimized to reduce acoustic diffraction effects (diffraction length is proportional to the square of the laser beam diameter) while simultaneously producing an optoacoustic (OA) signal of sufficient amplitude (inversely proportional to the square of the laser beam diameter). Thus, the laser beam diameter is chosen to be 1.5 mm for 3 mm thick samples.

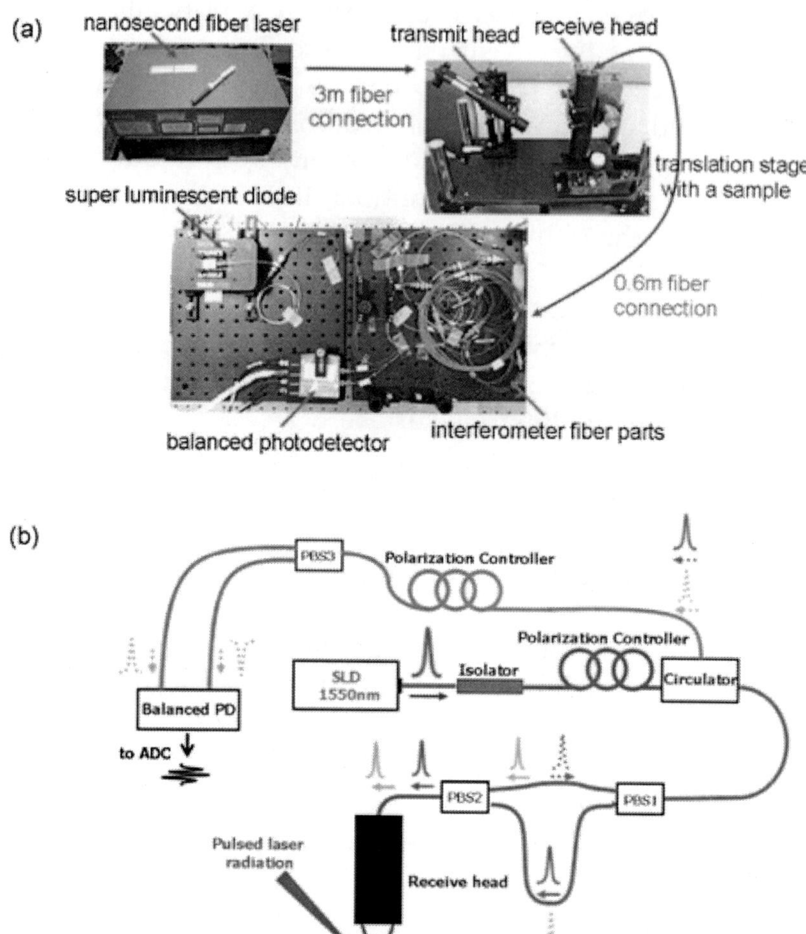

Figure-2 (a) Main components of the all-optical system for NDT&E of composite materials: compact very high repetition rate (up to 76 kHz) fiber laser is in the upper left corner; transmit head connected to the fiber laser with 3 m fiber bundle and receive head connected to the interferometer with PM fiber are in the upper right corner; photo of the modified fiber-optic Sagnac interferometer is in the lower center. (b) Operating principle of the interferometer: probe laser radiation is divided into two beams traveling by different length fibers to form a delay in the forward path; beam polarizations are exchanged after their reflections from the sample under study, forcing the beams to exchange paths back and thereby to interfere. Finally, scattered probe optical radiation is detected with the balanced photodetector.

The output fiber length is 3 m, making delivery of radiation very flexible. The laser source is also very flexible, permitting software control over several key parameters such as pulse duration, pulse energy, and pulse repetition rate. Pulse duration can be set to 10 ns, 30 ns, 60 ns, 120 ns or 250 ns. A 60 ns pulse duration is used to provide an US inspection bandwidth better than 10 MHz. The energy of laser pulses can be varied as well and the maximum is 0.53 mJ for the 60 ns pulse duration. Finally, the pulse repetition rate must be carefully selected based on the mechanical scan rate of the pump laser source. For example, a laser operating at an energy of 0.53 mJ at a 76 kHz pulse repetition rate would burn the sample immediately if it is fixed at a certain position, i.e. without scanning. The safe pulse repetition rate at this laser energy is determined by the heat release process, which for the class of samples used here is ~500 Hz with no sample scanning. However, the pulse repetition rate can be greatly increased when the sample under study moves, and is limited practically by the maximum translation speed of the scanner. To eliminate potential damage when the scanner was stationary, a USB switch controller (Numato, numato.com) was installed into the system to turn laser firing off immediately after sample translation terminated.

Incident pump-laser pulses are delivered at an oblique angle (~40° from the sample normal) that allows the probe-laser beam to be focused to the same point on the sample surface. The OA-generated pulse propagates through the sample, partially reflecting off of the composite structure and partially reflecting off of the back wall. All acoustic waves propagating back to the front wall of the sample are detected with the interferometer described below (Section 2.3). Radio frequency (RF) signals output from the interferometer are amplified in the frequency range of 1–10 MHz by the amplifier (Panametrics, Model 5072PR), digitized to 14 bits by the PCI Express3 ADC (GaGe, Model Razor Compuscope RZE-002-300, gage-applied.com) and transferred to the workstation (HP, Model Z820, hp.com) for further signal processing and display. The ADC is triggered by the output of a photodiode (Thorlabs, Model DET 10A, thorlabs.com) detecting a small fraction of the pump laser radiation coupled into it. This approach sets zero time as the moment when the OA pulse was generated at the sample surface and avoids any jitter induced by pulse-to-pulse instability of the laser.

Composite samples are positioned on a 2D translation stage for all imaging studies. The X-axis is driven by a stepper motor controller (Thorlabs, Model LNR50DD) with variable speed control up to 8 mm/s. The sample is moved continuously during scanning, where the maximum travel distance is 50 mm (maximum allowed travel distance for the stage). Position accuracy is also determined by the stage and is better than 2 μm. At each position during scanning, a digital RF A-scan is recorded for every laser firing. Each A-scan corresponds to the distribution of US transients scattered by the composite structure back to the detection point, i.e. along the Z-direction. Sample translation along the X-axis forms a B-scan image corresponding to the distribution of US scatterers in the XZ plane. Single A-scans in a B-scan are stitched together without any beamforming procedure or signal interpolation.

Translation of the sample in the Y-direction is performed manually to acquire a complete 3D data set characterizing the composite sample, and this data set can be used to form C-scan and M-scan movies. Software was developed to integrate this simple scanning system with the LU system and ensure proper synchronization of all components.

We note here that the scanning procedure is not in any way novel, and certainly does not represent the state of the art in composite materials testing. The scanners were selected primarily for their accuracy, not speed, and lower pulse repetition rates are used to match their scan speed. As will be discussed in some detail below, the ultimate scan speed of an LU system is limited by the maximum pulse repetition rate of the pump laser. For the 76 kHz maximum rate of the fiber laser used here, the ultimate scan rate of inspection is much higher than the 8 mm/s limit of the current system.

Fiber Optic Interferometer

The optical interferometer is a key component of all LU inspection systems. To meet the design goals presented in Section 1, we have developed a fiber optic, Sagnac-based interferometer with a balanced detection scheme. Below we summarize its main features. More detailed information on interferometer design, its sensitivity, and its overall imaging performance compared to that of a contact ultra-wide band PVDF detector, can be found in a recent paper from our group [24].

Probe Beam Source

A super luminescent diode (SLD, Thorlabs, Model SLD1550P-A40) is the source of probe radiation. It operates at a center wavelength of 1550 nm with a bandwidth of 60 nm. The center wavelength was chosen to match the communications range and, therefore, take advantage of the most recent developments in fiber-optic components and devices. The output power was proportional to the current applied to the SLD and can be varied up to 40 mW.

The probe laser is partially coherent (coherence length is ~40 μm). This feature is important because it removes nearly all parasitic interference inside the interferometer due to reflections between connections and, thus, dramatically decreases system noise. On the other hand, a 40 μm coherence length is many times larger than the displacement induced by the US transient at the sample surface.

The output probe radiation is polarized along the slow axis of a polarization maintaining (PM) fiber. An additional polarizer is attached to the SLD to ensure that all probe radiation is linearly polarized. A fiber isolator follows the polarizer to avoid any reflections back to the SLD which could damage the source.

Operating Principle

As with all interferometers, a signal is detected by interfering two independent optical beams (see Fig. 2b). The advantage of the Sagnac approach is that no reference beam is required. Both interfering beams come from the reflection of the probe by the sample surface, making the interferometer absolutely stable to any environmental vibrations and temperature fluctuations. In addition, this interferometer does not require feedback, a critically important issue for overall system stability, reliability, and speed. Furthermore, no reference arm makes the Sagnac interferometer relatively insensitive to the speckle structure of the optical probe beam reflected from a rough composite surface (see Fig. 1). Sample roughness, therefore, only influences the probe light power coupled into the interferometer.

To produce two independent beams, laser radiation initially linearly polarized along the slow axis is rotated by 45° using a polarization controller (Thorlabs, Model FPC020) and then divided into two interferometer arms with a polarization beam splitter (PBS1 in the drawing). These two fiber arms have different lengths (0.5 m and 10.5 m, respectively) so that two optical waves appear at the next polarization beam splitter (PBS2) with a fixed delay. This delay determines the maximum detectable frequency of US vibrations and can be easily adjusted by changing the longer fiber to a different length. Matching the fiber length to the desired frequency band (10 MHz in our case) maximizes detection sensitivity.

PBS2 combines the two delayed beams into one fiber, maintaining their polarizations. The collimator (receive head in Fig. 2b) finally focuses the probe radiation onto the sample surface. The design of the receive head is quite simple. It must focus probe radiation to the sample surface and couple backscattered radiation with maximum sensitivity back into the 8 μm PM fiber. Although simple in design, manufacturing of the head (CourierTronics, couriertronics.com) must be very precise. It contains two lenses, one collimates outgoing radiation from the fiber and the other, with a high NA of 0.5, focuses radiation onto the sample. There are a few additional components between the lenses: a wave separator (Altechna, altechna.com) to propagate 1550 nm radiation without distortion and block the pump 1064 nm laser radiation at the same time; and a high aperture (40 mm) quarter wave (QW) plate (Altechna) exchanging the polarization of the two interfering beams after their reflections from the sample surface.

Probe radiation reflected from the sample surface and coupled back to the fiber contains two delayed waves, similar to the conditions for incident illumination, but with exchanged polarizations. The wave that propagated initially through the short arm of the interferometer now propagates through the long arm on the way back, and vice versa for the second interfering beam. The two beams have no delay when they reach PBS1 except for the acoustic vibration induced phase difference, and they finally interfere after being converted to circular polarization states with a fiber polarization controller. The beams are split one last time with PBS3 to dramatically reduce interferometer noise and make the system insensitive to thermal lens [25] effects induced by the pumping laser. When the polarization controller (PBS3 in the drawing) is tuned to act as a perfect quarter wave plate, the two beams have opposite signs in the interference term at the two inputs of a balanced photodetector. Subtraction of the signals within the balanced photodiode (Thorlabs, Model PDB 420C-AC [26]) finally gives a photocurrent that is virtually insensitive to laser induced changes of the sample refractive index, and other stationary polarization insensitive noise. This makes it possible to work with very low light power reflected back by a rough sample surface. Note that the sensitivity of regular (non-avalanche) photodiodes is much less than that of avalanche ones, but they provide much better inherent dynamic range, better than 50 dB [27]. Overall, balanced detection provides sensitivity equivalent to that of avalanche photodiodes, but with much better dynamic range.

As noted above, the Sagnac scheme registers the difference between two surface displacements recorded at close time instants determined by the propagation delay in the long fiber arm relative to the short one. Thus, the interferometer output is proportional to vibration speed or acoustic pressure. This is another advantage over displacement-based interferometers commonly used in NDT&E applications since a derivative (i.e. high pass filter) operation is not required to remove low frequency displacement artifacts from raw spectrometer outputs. This high pass filtering operation, usually implemented in the digital domain, can reduce the SNR of the detected ultrasound signal.

To summarize the operating principle, when there is no pump laser impact at the composite surface, the displacement difference between the two interfering optical beams is zero, and the interferometer records nothing. When pulsed laser radiation generates an OA pulse, however, the interferometer detects the laser induced pressure signal, and all other pressure transients reflected back by the composite structure.

As mentioned above, the proposed Sagnac scheme consists of only fiber components. Most are polarization maintaining and virtually insensitive to both mechanical vibrations applied to the side of the fiber and fiber twisting. This makes interferometer design very flexible because fiber optical elements can be placed on any supporting surface, rather than needing to be solidly attached to an optical table. Once the interferometer has been tuned, there is no need for further tuning or any kind of feedback to stabilize operation. Overall, this is a very rugged design appropriate for typical field operations in the aerospace industry.

Finally, the last key feature of the Sagnac approach is that the detection bandwidth is not limited from below and is limited from above only by the difference in fiber arms of the interferometer determining the maximum operating frequency. Therefore, the maximum detectable frequency can be easily varied by changing the long arm fiber to another length, providing ultra wide-band US detection yet enabling optimal sensitivity over a specified frequency band. In the remainder of the studies reported below, the operating frequency band was chosen to span 1–10 MHz based on the acoustic properties and geometry of the composite materials under investigation.

Impulse response of the Sagnac detector

The optical detector's impulse response was measured by inputting an acoustic pulse with a much broader bandwidth than the detector itself and recording the resultant output. The interrogating pulse was created optoacoustically using a very strong optical absorber, permanent marker ink, painted on the surface of a PMMA plate. Fig. 3a presents the measured impulse response of the optical detector prior to any analog filtering. Clearly, the output has a smooth temporal profile with no sidelobes. Using the full width at half maximum amplitude, the pulse duration, Δt, is 90 ns. This can be easily converted into the equivalent spatial resolution along the propagation direction in a material using the speed of sound in that material.

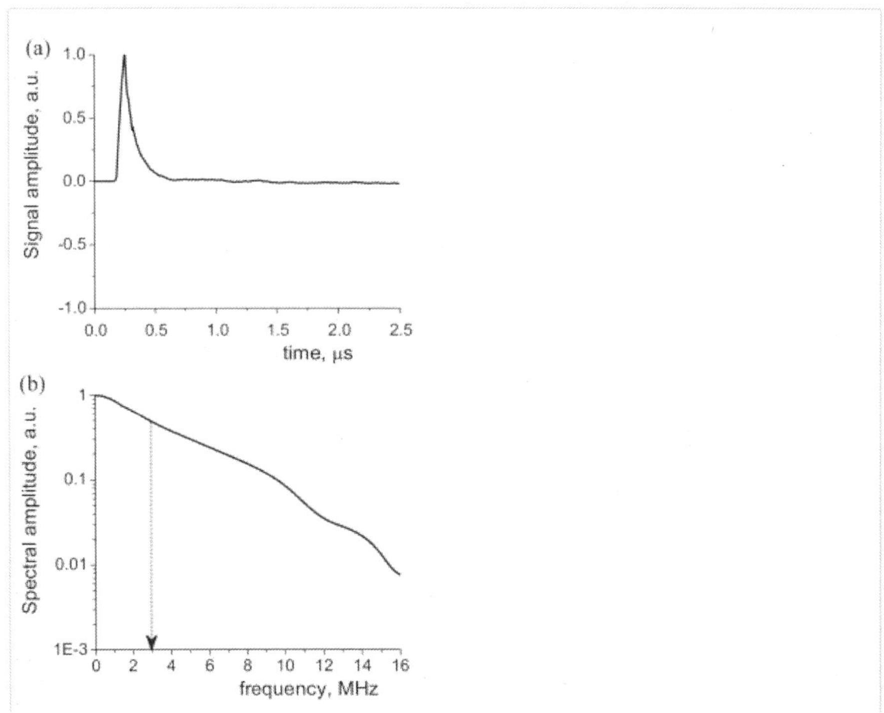

Figure-3 (a) Impulse response of the fiber-optical Sagnac detector; (b) corresponding spectra of the signal above. The characteristic frequency of the whole bandwidth signal is 2.9 MHz (at ½ level - dashed line in (b) but the impulse response remains compact at 95 ns duration).

The transfer function is the Fourier transform of the impulse response, as illustrated in Fig. 3b for the optical detector. Note that the upper frequency roll-off is determined primarily by the length of the long arm fiber in the interferometer and can be easily tuned by changing the length of that fiber. We have limited the acoustic spectrum to about 10 MHz since attenuation in the composite samples used here increases significantly at higher frequencies.

The detector bandwidth, as measured using the spectral width at half maximum amplitude, is 2.9 MHz but results in an impulse response only 95 ns in duration. A typical bipolar ultrasonic pulse produced by a broadband piezoelectric transducer operating at the same characteristic frequency would have a duration about equal to the inverse of the bandwidth, i.e. about 300 ns in this case. The tight impulse response of the optical detector is a significant advantage for NDT&E, typically producing a factor of three improvement in axial resolution compared to equivalent contact piezoelectric transducers.

RESULTS

Recorded A-Scan Signals And Their Processing

The reference signal illustrated in Fig. 3a was obtained from an optically flat surface and can be used directly for high resolution imaging without any additional signal processing. An inhomogeneous and rough surface, however, does not produce a stable low frequency signal. This is especially true when there are significant thermal lens effects at a particular point on the rough surface of a composite sample producing a very strong but very low frequency signal [23]. This signal can dominate the dynamic range of the ADC in the signal chain, thereby reducing the overall dynamic range of the system. To overcome this problem, the interferometer output is high pass (HP) filtered (using the 1 MHz Panametrics cutoff limit for the low frequency edge) prior to amplification and digitization. The resulting probe signal passed through the HP analog filter is illustrated in Fig. 4a with the dotted red curve.

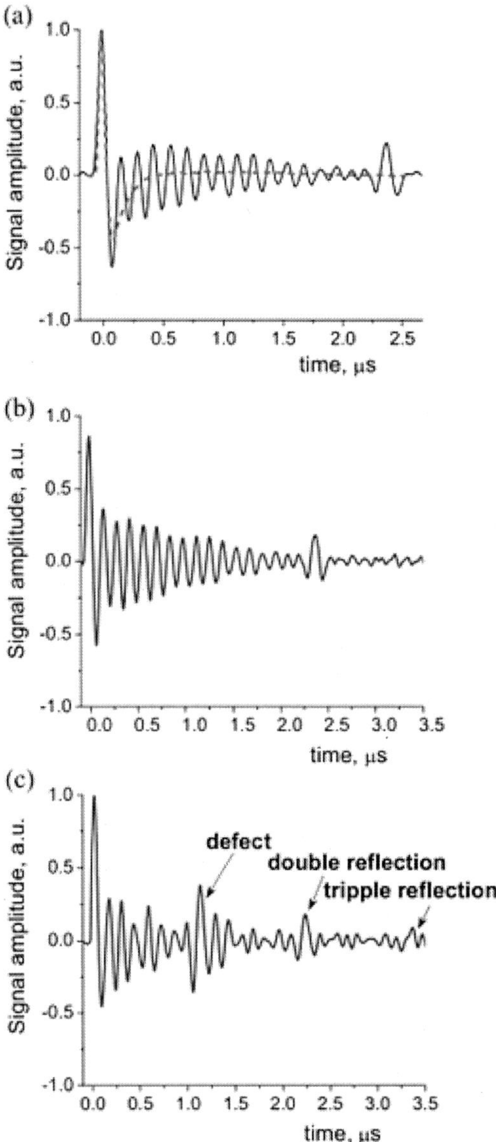

Figure-4 .(a) Single shot (without signal averaging) A-scan signal recorded for the composite sample UW-M (black line) and impulse response of the Sagnac interferometer PMMA after applying a bandpass (1 MHz – 10 MHz) filter (red line); (b) the same A-scan after deconvolution (according to Eqs. (1) and (2)) in a region free of defects; and (c) deconvolved A-scan in a region with an inclusion (brass foil).

Using the samples described in Section 2.1, a large collection of A-scans was recorded to further optimize the signal processing path for imaging applications. A typical A-scan signal recorded from a single pump laser firing (i.e. without any averaging) by the Sagnac detector and passed through HP analog filtering is plotted as the black curve in Fig. 4a. It corresponds to a region of the sample UW-M without inclusions. The first positive peak corresponds to the OA signal generated at the front sample surface due to absorption of pulsed laser radiation. The last peak signal arrives at the detection point after the generated signal's reflection from the back wall of the sample. All transients between these two represent reflections from inside the composite structure and form an oscillating periodic signal. Note that the composite samples under study consist of 19 layers which "produce" 18 corresponding US signal maxima.

Raw recorded signals are not ready to be imaged because they have an artificial negative tail following the front surface OA signal (black curve in Fig. 4a) resulting from HP filtering. To recover the full resolving power of the optical detector, a deconvolution procedure is applied to recorded A-scan signals to approach the "ideal" unipolar temporal profile [28] and [29] as much as possible.

Since the light absorption coefficient of the composite samples under study is more than 200 cm^{-1} (about 50 μm heat release depth), it is reasonable to assume that the OA signal generated at the front surface of the composite samples has a temporal profile identical to that of the one generated on the surface of the painted PMMA plate. Taking into account the sound speed in the composite of about 3000 m/s, we estimate that the spatial scale corresponding to US signal propagation during the laser pulse duration of 60 ns is equivalent to 180 μm. Thus, the profile of the laser-generated US signal is mostly determined by the laser pulse envelope. Differences in light absorption coefficients and Gruneizen parameters of PMMA and composite materials modulate PA signal amplitudes, but not signal profiles.

HP filtering by the Panametrics amplifier affects the Sagnac impulse response and the composite OA signal in precisely the same way. This means that for all frequencies outside the band of the very low frequency (below a few hundred kHz) instabilities of the composite OA signal, the only difference in its profile compared with the reference is due to scattering by sample heterogeneities. Thus, inverse filtration (or deconvolution) of the A-scan recorded for the composite sample with the reference OA signal can be represented in the frequency domain as:

$$S_{processed}(f) = \frac{S_{HP}(f)}{S_{Ref}(f)} \times Filter(f). \quad (1)$$

In Eq. (1), Sprocessed(f) is the spectrum of the resulting deconvolved signal, SHP(f) is the spectrum of the recorded A-scan after analog HP filtering with the Panametrics amplifier, and SRef(f) is the spectrum of the OA reference signal from the PMMA plate. Filter(f) is the spectrum of a bandpass filter designed to remove unwanted very low and very high frequencies. For the results presented below, this filter is defined as:

$$Filter(f) = \left(1 - \exp\left(-\left(\frac{f}{f_0}\right)^2\right)\right) \times \exp\left(-\left(\frac{f}{f_1}\right)^2 - \left(\frac{f}{f_2}\right)^4\right), \quad (2)$$

where f_0 = 100 kHz, f_1 = 10 MHz, f_2/f_1 = 1.2. The resulting processed signal is produced by inverse Fourier transformation of Sprocessed(f). The result of this processing for the signal of Fig. 4a is presented in Fig. 4b. Visually perfect reconstruction cannot guarantee full signal recovery, but inverse filtering clearly reduces the negative slope in composite signals and enables full utilization of the dynamic range of the ADC.

Fig. 4c shows a processed Sagnac recording (after applying deconvolution) for a region containing a thin brass foil inclusion in the composite structure. The US reflection from the inclusion is clearly seen, as well as high-order reverberations between the defect and the front surface. The artificial negative slope is well compensated with deconvolution filtering. This procedure is applied to all A-lines used to create the B-scan images shown below.

B-Scan Images

Single Shot Images

Fig. 5 illustrates B-scan images obtained within the defect region of the sample UW-M. Each B-scan consists of 450 A-scans, normalized by their amplitudes at the front surface. Normalization was needed because the light absorption coefficient of the composite sample, local sample roughness, and small displacements of the Sagnac focal point from the sample surface can change from point to point. All of these factors influence only the measured signal amplitude, however, and do not appreciably change its profile.

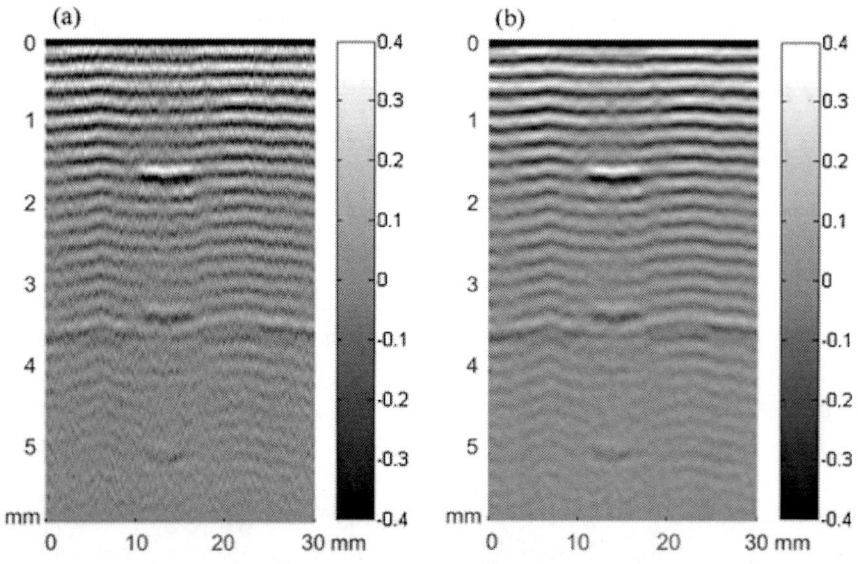

Figure-5 Typical B-scan images obtained with Sagnac interferometer (a) for single-shot regime and (b) after application of 10 A-scan moving average. All A-scan signals were normalized by their maxima prior to forming a B-scan image. Color bars show the distribution of signal amplitude over a linear gray-scale.

The sample was moved at a constant speed of 1 mm/s over a distance of 30 mm in the lateral direction, corresponding to a 66 μm spacing between A-scans. The laser pulse repetition rate was 15 Hz, but could be increased to 1 kHz keeping the same sample translation speed, and ultimately to 76 kHz for applications requiring very fast translation. A single shot (i.e. one laser firing per A-line) image obtained with the Sagnac is presented in Fig. 5a. As seen, the regular layered composite structure is visualized very well along with the well-defined inclusion.

As is evident from Fig. 5, the sensitivity of the Sagnac detector is sufficient for non-contact imaging of the composite materials used in this study even in a single shot regime. The signal to noise ratio (SNR), computed as the ratio of the front wall OA signal to the noise pressure, is about 26 dB, or 14 dB if referenced to the back wall signal. Furthermore, since the pump fiber laser can fire at a very high repetition rate (76 kHz for the current laser), scanning can be performed very quickly while additional signal averaging can be applied. Because the lateral resolution generally cannot be better than the pump laser beam diameter, 1.5 mm in our case, and the scanning step is only 66 μm, a simple moving average filter can be applied in the translation direction. Such processing does not substantially degrade lateral resolution if the averaging region is smaller than 750 μm (half of the laser beam diameter). The Sagnac based B-scan resulting from a moving average of 10 adjacent RF A-scans (i.e. 660 μm averaging window) is illustrated inFig. 5b. As seen, the SNR is greatly (by $\sqrt{10}$ times, or 10 dB) increased with this simple processing without any apparent change in spatial resolution. Note that moving average filtering can also be applied in two dimensions for three-dimensional data acquisition.

Different Defect Locations

As noted above, samples with inclusions placed at different depths were studied. The simplest case in US defectoscopy usually corresponds to a defect located far from sample surfaces (Fig. 5). In contrast, if the defect is located close to the front sample surface, detection and precise localization is sometimes challenging due to insufficient in-depth resolution or a dead zone for US detection in the very near field.

Fig. 6a illustrates the case where a thin brass foil inclusion is positioned only one layer, or 190 μm, below the sample surface. In spite of the modest 2.9 MHz probe signal bandwidth, the defect is imaged very well, demonstrating that the Sagnac in-depth resolution is better than one composite layer. In addition, multiple reverberations of the probe OA signal between the defect and the front sample surface are observed. An important feature of an ultra-wideband LU system is that the temporal profile of the probe OA signal coincides with the signal envelope. That is, not only impedance discontinuities can be detected, but the sign of that discontinuity can be identified as well. Therefore, defects with impedance higher than that of the composite should be darker in the B-scan whereas defects with lower acoustic impedance should be brighter for our display format. We did not observe this in Fig. 6a, however, where a white strip above the black colored brass foil is quite evident. In later discussions with the material fabricator we discovered that the brass had not adhered well to the composite matrix for this sample, presumably producing an air gap leading to the bright signal observed in the B-scan.

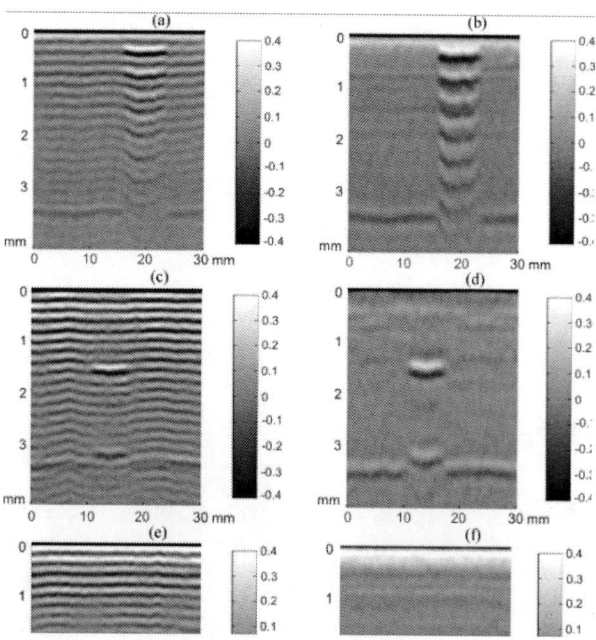

Figure-6 Typical B-scan images obtained with the fiber-optic Sagnac interferometer for different defect locations: (a) close to the front surface, (c) in the middle of the sample, (e) close to the back wall. Panels (b), (d) and (f) are low-pass (Eq. (3)) filtered versions of the images in the left column. Low-pass filtering greatly reduces signals from the regular composite structure and enhances the visibility of surface-like reflections from inclusions and sample boundaries. A moving average over 10 adjacent A-scan was applied to the data set. All A-scan signals were normalized by their maxima prior to forming a B-scan image. Color bars show the distribution of signal amplitude over a linear grayscale.

Figure. 6c presents the B-scan for the case of a brass foil positioned approximately in the middle of the sample. The signal close to the back sample wall is the second reverberation of the probe OA signal between the defect and the front sample surface. The third reverberation is also clearly seen (Fig. 5b) and not shown here. Higher order reverberations can also be observed if the sampling volume is extended.

The case where the brass foil is located a few layers above the composite back wall is illustrated in Fig. 6d. As seen, it is very clear distinguished from the back sample wall, as the signal reflection from it is totally blocked by the defect. Most probably, this defect was not also perfectly embedded into the composite matrix and a thin air cavity produces the bright strip beyond the brass inclusion.

Low-Pass Image Filtration

For B-scans using the full signal bandwidth, signals associated with the regular composite structure as well as surface-like reflection are well visualized. Note that sometimes signals from an inclusion can be masked by the strong US transients coming from the heterogeneous sample structure. In this section we present the results of a simple processing method to minimize signals from the regular composite structure, enhancing the visibility of surface-like reflections from inclusions and sample surfaces.

Since the spectrum of LU signals is ultra wideband, it can be used to identify signal harmonics characterizing different material structure components. Fig. 7 shows a typical spectrum of an A-scan signal recorded in an area free of defects. A strong narrow pole centered at a frequency of 7.1 MHz is associated with the strongly periodic composite structure. By eliminating this component, signals from the periodic composite structure are deemphasized. A low-pass filter can perform this task but it must be designed to maintain as much of the signal bandwidth as possible and produce a unipolar impulse response. The simplest operation to meet these requirements is to apply a filter to the signal in addition to the filter function described by Eq. (2), changing the characteristic frequencies f_1 and f_1 so that tof_1LP = 5 MHz and f_2LP/f_1LP = 1.1:

$$LP\_Filter(f) = \left(1 - \exp\left(-\left(\frac{f}{f_0}\right)^2\right)\right) \times \exp\left(-\left(\frac{f}{f_{1LP}}\right)^2 - \left(\frac{f}{f_{2LP}}\right)^4\right), \quad (3)$$

as illustrated in Fig. 7 by the red curve.

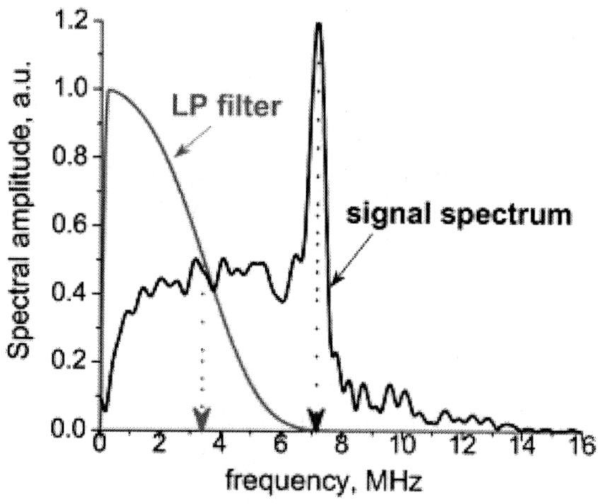

Figure-7 Frequency spectrum of an A-scan signal (black curve). A strong narrow peak at 7.2 MHz corresponds to the regular periodic structure of the composite sample. The low-pass filter function shown in red reduces the structure component of the signal.

The results of applying the LP filter to the data used to produce the left column of B-scans in Fig. 6 are presented on the right side of the same figure. Clearly, the strongly periodic structure signal is mainly removed from all B-scans, and US signal reflections corresponding to the defects are seen more clearly. In addition, large scale structure imperfections which were totally invisible in full bandwidth images are recognizable, demonstrating the power of the wideband LU technique. On the other hand, the axial resolution is degraded, resulting in "blurring" of defect signals and loss of high frequency information. More robust methods that can minimize the periodic structure signal while maintaining more of the overall signal bandwidth will be the subject of future studies.

LU Imaging With Very Rough Sample Surface

As noted in Section 1, composite sample roughness is a serious factor dramatically affecting the efficiency of US detection with optical interferometers. In this study, composite samples had surfaces with very different roughnesses. The front surface of all samples had a roughness matching that of factory-made components used in the aerospace industry (Fig. 1a). In contrast, the back surface of all composite samples (see Fig. 1b and c) was extremely rough.

Fig. 8a presents a typical distribution of recorded OA signal amplitudes versus scan distance over a 30 mm segment for the case where the optical pump/probe beams are delivered to the front surface. Small scale amplitude variations are determined by sample roughness whereas long scale variations are related to surface non-flatness altering the position of the probe beam focal spot. Clearly, the amplitude does not change dramatically, and these variations are easily compensated with signal normalization prior to imaging. Signals recorded for the case where optical pump/probe beams are delivered to the back surface are much more variable, as illustrated in the blue curve of Fig. 8b where almost two thirds of detected amplitudes are below the noise level. A full bandwidth B-scan image for this case is presented in Fig. 9a. The defect can be seen, but the image quality is not comparable to that demonstrated earlier. Even with a 30 point moving average filter across A-scans (Fig. 9b), image quality is not fully recovered.

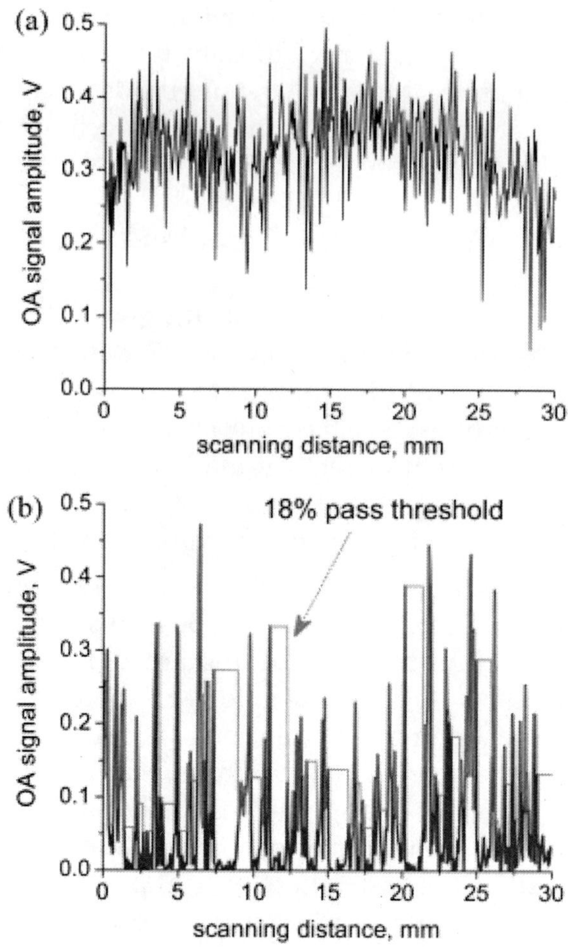

Figure-8 Typical amplitude dependence of the A-scan signal over a 30 mm scan distance for (a) regular front surface and (b) extremely rough back surface (blue curve). The red curve in panel (b) represents an 18% threshold cutoff below which A-scans are ignored and replaced with adjacent ones in B-scan images.

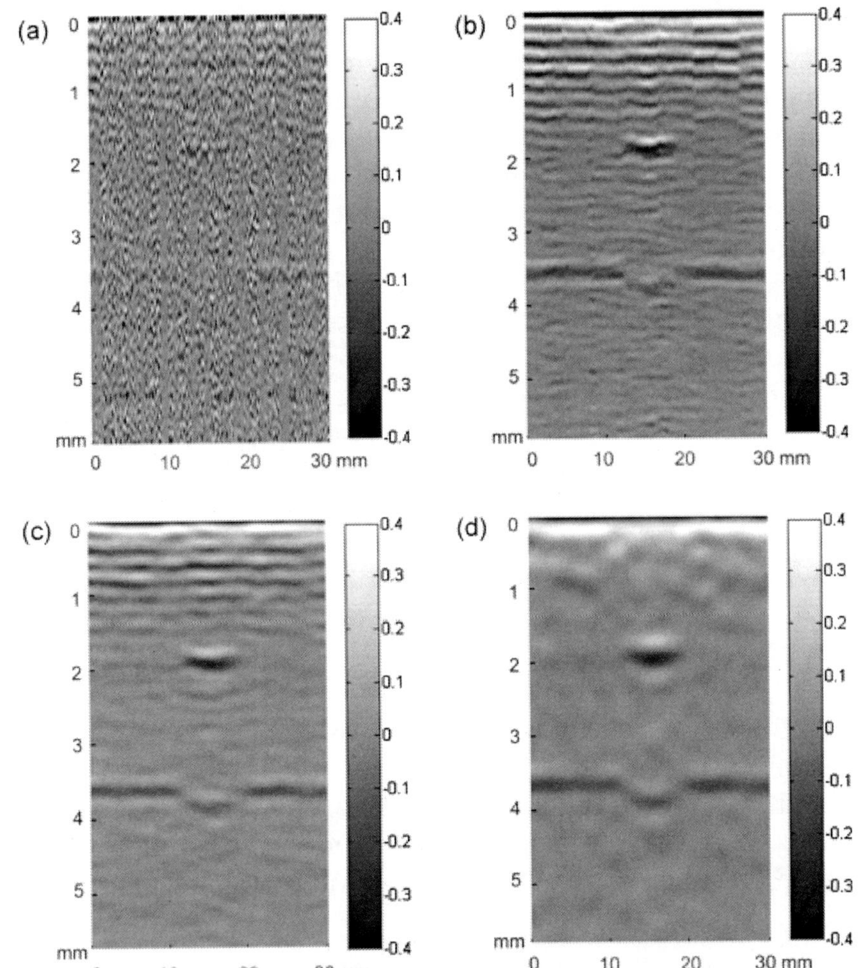

Figure-9 B-scan image acquired with an extremely rough surface (shown in Fig. 1): (a) single shot B-scan; (b) B-scan after application of 30 A-scan moving average; (c) B-scan after application of the 18% cutoff filter and 30 A-scan moving average; (d) B-scan after the 18% cutoff filter, 30 A-scan moving average, and LP filtering. All A-scan signals were normalized by their maxima prior to forming a B-scan image. Color bars show the distribution of signal amplitude over a linear gray-scale.

Although signal quality is significantly degraded for a very rough surface, some simple signal processing can recover most of the information require to make a useable B-scan image with the Sagnac detector even for the very rough surfaces considered here. Roughly one third of the recordings presented in Fig. 8b are above the noise level. An image can be reconstructed using only these signals and ignoring the rest. Using a threshold defined as 18% of the maximum recorded amplitude in the scan, a value derived empirically for this case, an A-scan will either contribute to the B-scan image or be replaced by its nearest neighbor that is above the threshold, as illustrated by the red curve in Fig. 8b. We note that the 18% threshold level was chosen empirically, but that the results presented below are not highly sensitive to the specific choice of this threshold level. B-scans reconstructed with this procedure are illustrated in Fig. 9c for whole bandwidth signals, and in Fig. 9d for LP filtered signals. Clearly, image quality is degraded compared to that in Fig. 6c and d. Nevertheless, the processed image is of sufficient quality to plainly identify the defect. We note again that these results were obtained with an extremely rough surface that would be nearly impossible to probe with speckle sensitive interferometers.

Complete Us Data Set

As mentioned above, multiple B-scan images were recorded with a step of 0.5 mm in the Y direction. A total of 360 B-scans were recorded to form a 3D data set for all samples under study. These data can be displayed using multiple formats. In Movies 1 and 2, these are presented to illustrate the overall imaging performance of the all-optical, non-contact system.

Each frame in the movie presented in Fig. 9 represents a B-scan slice through the full 3D data set. The upper panel shows the current position of the B-scan imaging plane relative to significant structures within the sample. B-scans movies are shown over the LP filtered (upper B-scan) and whole bandwidth (lower B-scan) formats described earlier for static images (see Fig. 6). Defects of all types (thin brass foil, tape and polymer) are very well visualized. The brass foil definitely exhibits the strongest reflected signals, which are even seen as double and triple reflections in Fig. 5. However, the thin tape inclusion is also well visualized with the non-contact, all-optical system.

When information about the regular composite structure is not required for evaluation, LP filtered images more clearly detect defect locations since periodic signals from the structure that could mask these reflections are virtually eliminated. We note also that artificial flashes appear from time to time in these movies. They correspond to A-scans recorded at points with deep surface scratches, cavities, or bumps that dramatically reduce the magnitude of reflected light that can be delivered to the interferometer. These bad A-scans can be easily identified and removed from B-scans using the processing described above to handle signals from the very rough back surface. In these movies, however, we deliberately kept all raw signals to demonstrate the overall robustness of our LU system.

Movie 1 shows the 3D data set where each frame represents an independent XY plane. In other words, each frame presents the image of heterogeneities in a plane parallel to the front surface (C-scan). There were 800 frames separated by 15 μm, corresponding to each sampling point of the ADC. Using the same display format as the B-scan movies, C-scan movies present both whole bandwidth and LF filtered images. This format can reveal composite structure layer by layer, with the resolution determined by the detection bandwidth. This movie clearly captures the periodic layer pattern of composites and also shows the depth and lateral extent of all defects. It is interesting that LP filtered C-scans clearly exhibit the lateral braiding of composite structure in the near field where the lateral resolution is primarily determined by the detection spot diameter rather than the diameter of the pumping laser beam.

DISCUSSION

Detection sensitivity is a key parameter for remote US measurements, because the sensitivity of optical interferometers is usually much smaller than that of piezoelectric transducers. In a companion study [24], we compared contact US measurements using an ultra-wide band PVDF detector made from a 28 μm PVDF film to equivalent measurements using the Sagnac detector. The NEP of the Sagnac scheme was better than 400 Pa. It is about 40 times worse than that of the PVDF transducer (the diameter of the sensitive area is 6 mm) operating in the open-circuit regime [27]. Note, that the sensitivity of wide-band PVDF transducers operating in the open-circuit regime is proportional to the capacitance of the piezo-element [27], i.e. to the surface area. However, wide-aperture transducers are good for inspection of quite thin samples when propagation of ultrasound can be considered in the near field. For thicker boards, spatial averaging over the transducer aperture blurs the image with reduced lateral resolution. Our 7.5 μm diameter non-contact detector is equivalent in SNR to a 1 mm diameter PVDF detector, representing the state of the art in contact wideband detection. Conversely, increasing the detection spot to a few mm in the Sagnac detector would improve SNR for a mirror-like surface, but it is not practical for rough and non-flat composite surfaces.

The absolute sensitivity of any detection scheme can be quantified by comparing the NEP to that of an ideal detector of the same bandwidth and aperture limited only by Johnson–Nyquist (i.e. thermal) noise. The noise power of the Sagnac detector is only 17 times higher than that of an equivalent ideal acoustic detector, representing a detector system noise figure of 12.3 dB (i.e. noise level above the Nyquist thermal noise limit, the ultimate value which cannot be overcome). Moreover, this figure overestimates the noise figure of the interferometer itself since it includes all of the electronics in the signal path. Such a low system noise figure for optical detection from rough composite surfaces reflecting less than 1% of incident light over a 7.5 μm diameter detection aperture should enable non-contact LU inspection systems with exquisite sensitivity and spatial resolution.

The Sagnac approach presented here is particularly appropriate for LU systems performing NDT&E studies on composite materials used in the aerospace industry because of its sensitivity and robustness. The interferometer design overcomes, in part, the limitations of previous ones. A number of issues were addressed in our design to reach the performance level described above. The interferometer is assembled from fiber-optic devices and includes a balanced detection scheme to improve sensitivity. Most components are custom designed and leverage the most recent advances in optical sources and fiber-optic components developed for the optical communications industry. A low coherence SLD in the interferometer greatly minimized the effects of destructive interference within the interferometer itself. We have also optimized all optical elements by maximizing transmission. Thus, using two broadband interfering beams with high intensities reflected from one surface maximizes constructive interference and minimizes noise. Balanced detection also greatly minimizes the effects of low frequency noise from thermal lens effects. Overall, this approach produced a robust interferometer operating at high light intensities but with minimal parasitic noise. The optical fiber providing the delay in the Sagnac was chosen to be 10.5 m to maximize detection sensitivity over a frequency band of ~10 MHz. Note, that the fiber can be easily switched to another length in this design to maximize transmission in the desired frequency band.

The main problem with LU inspection of composite samples is that the sample surface is quite rough. In this work, we presented results acquired from composite samples with different surface roughness. For surfaces matching normal manufactured quality, the amplitude of detected US signals has a standard deviation of about 20%, as illustrated in Fig. 8a. This variation, however, does not affect the temporal profile of recorded signals and simply changes overall amplitude. Furthermore, it was very surprising that Sagnac detection still works (see Fig. 9) when sample roughness is extremely high with signal amplitude deviations more than 130% of the mean value (see Fig. 1 and Fig. 8). Nevertheless, simply ignoring bad events (in our case, recorded signals with amplitude less than 18% of the maximum) and applying a moving average across 30 adjacent A-scans produced B-scan of sufficient quality to clearly identify a known defect. Future studies will focus on more sophisticated signal processing methods to better recover image quality under these circumstances. Nevertheless, the example presented here of non-contact imaging from a very rough sample surface demonstrates the power, stability, and overall sensitivity of the differential Sagnac approach that is relatively insensitive to speckle and does not require a reference interferometer arm.

A possible challenge of this system for in-field applications is a relatively small depth of field. Since the Sagnac detector is focused, it receives backscattered light only from the focal area of the detector. When the surface is not flat, i.e. surface height is greater than 0.2 mm over the scanning range in the lateral direction, the SNR will degrade, potentially requiring an additional self-positioning tool for the vertical coordinate. Another way to improve the depth of field is increasing the focal length of the lens in the receive head, but decreasing the NA of the lens will definitely affect the amount of collected light and thereby the resulting sensitivity. Optimizing the depth of field for a certain surface roughness will be the focus of future studies.

In future studies, we also plan to exploit the Sagnac detector to evaluate composite material properties using a fully non-contact approach. For example, the ultra-wide bandwidth of the present system can potentially be exploited for non-destructive imaging of material porosity. A high-sensitivity, point-like US detector can also be used in non-contact detection of surface acoustic waves, potentially enabling a wide range of applications [30]. Finally, the detector's small size and high sensitivity may be appropriate for a number of biomedical problems.

CONCLUSIONS

The performance of a non-contact, all-optical, compact, inexpensive LU system for NDT&E of composite materials has been demonstrated on samples containing a wide range of defects. High resolution images of both the inherent composite structure as well as all inclusions within the sample clearly showed the efficacy of the system. This system was also tested for US signal detection from extremely rough surfaces, and surprisingly showed excellent performance using a simple data processing algorithm. The key component of the system is a modified Sagnac-based fiber optical balanced interferometer, which is quite robust for practical applications – it does not require a reference arm; detects acoustic pressure instead of displacement; does not need adjustment and stabilization; and can be mounted on any surface. It is also quite insensitive to parasitic interferences within the interferometer itself because a short 40 μm coherence length SLD is used as the optical source. The interferometer exhibits very good sensitivity in ultra wide-band US signal detection – noise equivalent pressure is about 400 Pa in the frequency band of 1–10 MHz for US signal detection from rough composite surfaces, representing a noise source only 12.3 dB higher than that of an ideal acoustic detector operating over the same bandwidth and aperture. To our knowledge, this is the best reported sensitivity for a non-contact ultrasonic detector of this dimension.

ACKNOWLEDGMENTS

The work reported here was supported by a contract between

the Boeing Company and the University of Washington (project# 66-1915), a grant from the Joint Center for Aerospace Technology Innovation (project# 06-1041), and the Department of Bioengineering at the University of Washington. We thank Dick Bossi, Jeff Kollgaard, Bill Motzer, and Jill Bingham at the Boeing Company for help with nearly every aspect of this project, and especially for supplying the composite samples.

REFERENCES

1. J.-P. Monchalin IEEE Trans Ultrason Ferroelectr Freq Control - UFFC, 33 (1985), pp. 485–499.
2. C.B. Scruby, L.E. Drain **Laser-ultrasonics: techniques and applications** Adam Hilger, Bristol, UK (1990).
3. R.J. Dewhurst, Q. Shan Meas Sci Technol, 10 (1999), pp. R139–R168 View Record in Scopus.
4. H. Li, L. Wang Phys Med Biol, 54 (2009), pp. R59–R97.
5. J.-P. Monchalin Appl Phys Lett, 47 (1985), pp. 14–16
6. A. Blouin, J.-P. Monchalin Appl Phys Lett, 65 (1994), pp. 932–934
7. A. Blouin et al. **Rev prog** D.O. Thompson, D.E. Chimenti (Eds.), QNDE 26, Plenum, New York (2007), pp. 193–200.
8. Blouin A, et al., US Patent# 6,813,951 B2 (2004).
9. B. Culshaw Meas Sci Technol, 17 (2006), pp. R1–R16.
10. T.S. Jang, S.S. Lee, B. Kwon, W.J. Lee, J.J. Lee IEEE Trans Ultrason Ferroelectr Freq Control, 49 (2002), pp. 767–775.
11. T.S. Jang, J.J. Lee, D.J. Yoon, S.S. Lee Ultrasonics, 40 (2002), pp. 803–807
12. T. Tachizaki, T. Muroya, O. Matsuda, Y. Sugawara, D.H. Hurley, O.B. Wright Rev Sci Instrum, 77 (2006), pp. 043713-1–043713-12.
13. P.A. Fomitchov, S. Krishnaswamy, J.D. Achenbachm Opt Laser Technol, 29 (1997), pp. 333–338.
14. T. Buma, M. O'Donnell Appl Phys Lett, 85 (2004), pp. 6045–6047.
15. C.-Y. Chao, S. Ashkenazi, S.-W. Huang, M. O'Donnell, L.J. Guo IEEE Trans Ultrason Ferroelect Freq Control, 54 (2007), pp. 957–965.
16. B.-Y. Hsieh, S.-L. Chen, T. Ling, L.J. Guo, P.-C. Li Opt Express, 20 (2012), pp. 1588–1596
17. T. Ling, S.-L. Chen, L.J. Guo Opt Express, 19 (2011), pp. 861–869
18. A. Rosenthal, D. Razansky, V. NtziachristosOpt Express, 20 (2012), pp. 19016–19029
19. B.F. Pouet, R.K. Ing, S. Krishnaswamy, D. Royer Appl Phys Lett, 69 (1996), pp. 3782–3784.

20. A.A. Kamshilin, A.I. Grachev Appl Phys Lett, 81 (2002), pp. 2923–2925.
21. S. Zamiri, B. Reitinger, E. Portenkirchner, T. Berer, E. Font-Sanchis, P. Burgholzer et al.Appl Phys B (2013) http://dx.doi.org/10.1007/s00340-013-5554-7
22. G. Rousseau, B. Gauthier, A. Blouin, J.-P. Monchalin J Biomed Opt, 17 (2012), pp. 061217-1–061217-7.
23. S.M. Rytov, Yu.A. Kravtsov, V.I. Tatarskii **Principles of statistical radiophysics** Springer-Verlag; Berlin, Berlin (1989).
24. I.M. Pelivanov, T. Buma, J. Xia, C.-W. Wei, M. O'Donnell JAP (2014) (accepted for publication, Article no. JR13-11986R).
25. N.G.C. Astrath, L.C. Malacarne1, G.V.B. Lukasievicz, M.P. Belancon, M.L. Baesso, P.R. Joshi et al. J Appl Phys, 107 (2010), pp. 083512-1–083512-6.
26. http://www.thorlabs.us/Thorcat/21600/PDB420C-Manual.pdf.
27. A.A. Oraevsky, A.A. Karabutov Proc SPIE 3916, biomedical optoacoustics (2000), pp. 228–239.
28. V.E. Gusev, A.A. Karabutov **Laser optoacoustics** AIP, New York (1993)
29. A.A. Karabutov, E.V. Savateeva, N.B. Podymova, A.A. Oraevsky J Appl Phys, 87 (2000), p. 2003.
30. P. Hess Phys Today, 55 (2002), pp. 42–47.

Chapter 5

CLAY/POLYMER COMPOSITES: THE STORY

Kay Hamacher

School of Biomedical and Natural Sciences, Nottingham Trent University, Clifton Campus, Clifton Lane, Nottingham, NG11 8NS, UK

Clay/polymer nanocomposites offer tremendous improvement in a wide range of physical and engineering properties for polymers with low filler loading. This technology can now be applied commercially and has received great attention in recent years. The major development in this field has been carried out over the last one and half decades. The progress, advantages, limitations, and current problems will be discussed in this review. So far, significant progress has been made in the development of synthetic methods, application to engineering polymers, and the investigation of major engineering properties. However, we are far from the end of the tunnel in terms of understanding the mechanisms of the enhancement effect in nanocomposites.

The nanocomposite approach has advantages over traditional fiber reinforced composites in the low filler loading range. Despite this, the market for high-performance fiber reinforced composites with high fiber volume fractions has not been affected by these developments.

The mystery of the 'nano-world' has been progressively exposed in recently years. The nanometer scale is simply a range between micro and molecular dimensions. The sciences in these two dimensional ranges have been well explored by materials scientists and chemists. Materials science and chemistry are often engaged in research on the nanometer scale, for example, the dimensions of crystal structures. The well-known nanometer-scale technologies developed within materials science and chemistry in the past may not be reasonably regarded as nanotechnology. The real interest in nanotechnology is to create revolutionary properties and functions by tailoring materials and designing devices on the nanometer scale.

Clay/polymer nanocomposites are a typical example of nanotechnology. This class of material uses smectite-type clays, such as hectorite, montmorillonite, and synthetic mica, as fillers to enhance the properties of polymers. Smectite-type clays have a layered structure. Each layer is constructed from tetrahedrally coordinated Si atoms fused into an edge-shared octahedral plane of either $Al(OH)_3$ or $Mg(OH)_2$. According to the nature of the bonding between these atoms, the layers should exhibit excellent mechanical properties parallel to the layer direction. However, the exact mechanical properties of the layers are not yet known. It has been estimated from recent modeling work that the Young's modulus in the layer direction is 50 to 400 times higher than that of a typical polymer1[, 2, 3, 4 and 5]. The layers have a high aspect ratio and each one is approximately 1 nm thick, while the diameter may vary from 30 nm to several microns or larger[6]. Hundreds or thousands of these layers are stacked together with weak van der Waals forces to form a clay particle. With such a configuration, it is possible to tailor clays into various different structures in a polymer.

In the past, the major interest in using clays for polymer enhancement was to break down clay particle aggregates into individual particles to form micro-sized filler reinforced polymers, as shown in Fig. 1. It can be imagined that the excellent mechanical properties of each individual layer in clay particles cannot function effectively in such a system. The weak interlayer bonding may act as damage initiation sites in applications. It is common to use high clay loading to achieve adequate improvement of the modulus, while strength and toughness of the polymer are reduced.

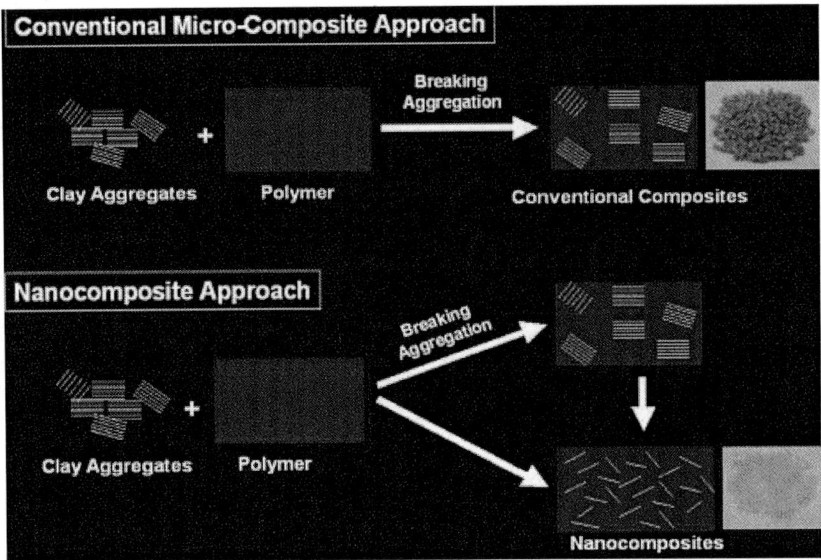

Figure 1. The different principles applied to the fabrication of conventional micro- and nano-composites.

The principle used in clay/polymer nanocomposites is to separate not only clay aggregates but also individual silicate layers in a polymer, as illustrated schematically in Fig. 1. By doing this, the excellent mechanical properties of the individual clay layers can function effectively, while the number of reinforcing components also increases dramatically because each clay particle contains hundreds or thousands of layers. As a consequence, a wide range of engineering properties can be significantly improved with a low level of filler loading, typically less than 5 wt%. At such a low loading level, polymers such as nylon-6 show an increase in Young's modulus of 103%, in tensile strength of 49%, and in heat distortion temperature of 146%[7]. Other improved physical and engineering properties include fire retardancy8 and 9, barrier resistance10, 11 and 12, and ion conductivity13 and 14.

Another advantage of clay/polymer nanocomposites is that the optical properties of the polymer are not significantly affected. The thickness of individual clay layers is much smaller than the wavelength of visible light so that well exfoliated clay/polymer nanocomposites should be optically clear. The images of micro- and nano-composites shown in Fig. 1 were produced using the same clay and polypropylene mixture and applying a rapid cooling processes to minimize the crystallization effect. The conventional microcomposites appear brown and opaque, while the nanocomposites are almost transparent.

It is clear from this evidence that clay/polymer nanocomposites are a good demonstration of nanotechnology. By tailoring the clay structure in polymers on the nanometer scale, novel material properties have been found. Another interest in developing clay/polymer nanocomposites is that the technology can be applied immediately for commercial applications, while most other nanotechnologies are still in the concept and proving stage.

The first commercial application of these materials was the use of clay/nylon-6 nanocomposites as timing belt covers for Toyota cars, in collaboration with Ube in 1991[15]. Shortly after this, Unitika introduced nylon-6 nanocomposites for engine covers on Mitsubishi's GDI engines[15]. In August 2001, General Motors and Basell announced the application of clay/polyolefin nanocomposites as a step assistant component for GMC Safari and Chevrolet Astro vans[16]. This was followed by the application of these nanocomposites in the doors of Chevrolet Impalas[16]. More recently, Noble Polymers has developed clay/polypropylene nanocomposites for structural seat backs in the Honda Acura[17], while Ube is developing clay/nylon-12 nanocomposites for automotive fuel lines and fuel system components.

In addition to automotive applications, clay/polymer nanocomposites have been used to improve barrier resistance in beverage applications. Alcoa CSI has applied multilayer clay/polymer nanocomposites as barrier liner materials for enclosure applications[18]. Honeywell has developed commercial clay/nylon-6 nanocomposite products, Aegis™ NC resin, for drink packaging applications[19]. More recently, Mitsubishi Gas Chemical and Nanocor have codeveloped Nylon-MXD6 nanocomposites for multilayered polyethylene terephthalate (PET) bottle applications[20].

THE HISTORY OF CLAY/POLYMER NANOCOMPOSITES

Although research on clay/polymer intercalation can be traced back to before the 1980s[21, 22, 23, 24 and 25], these developments should not be taken into account in the history of clay/polymer nanocomposites as the work did not result in dramatic improvement in the physical and engineering properties of polymers. The era of clay/polymer nanotechnology can truly be said to have begun with Toyota›s work on the exfoliation of clay in nylon-6 in the latter part of the 1980s and the beginning of the 1990s26, [27 and 28]. It was this work that demonstrated a significant improvement in a wide range of engineering properties by reinforcing polymers with clay on the nanometer scale29 [and 30]. Since then, extensive research in this field has been carried out globally. At present, development has been widened into almost every engineering polymer including polypropylene (PP)[31], polyethylene[32], polystyrene[33], polyvinylchloride[34], acrylonitrile butadiene styrene (ABS) polymer[35], polymethylmethacrylate[36], PET[37], ethylene-vinyl acetate copolymer (EVA)[38], polyacrylonitrile[39], polycarbonate[40], polyethylene oxide (PEO)[41], epoxy resin[42], polyimide[43], polylactide[44], polycaprolactone[45], phenolic resin[46], poly p-phenylene vinylene[47], polypyrrole[48], rubber[49], starch[50], polyurethane[51], and polyvinylpyridine (PVP)[52].

FABRICATION TECHNOLOGY

The ultimate aim of fabricating clay/polymer nanocomposites is to separate and disperse individual clay layers in a polymer. The applied strategy depends on the compatibility of the clay and polymer to be used. This determines if pretreatment of the clays and polymers is necessary before intercalation. If the surface of the silicate layers in the clays is compatible with the polymer, direct intercalation between the two can occur without the need for pretreatment. This is the case with water-soluble polymers such as PEO and PVP. These polymers and the surface of silicate layers are all hydrophilic. The dipolar or van der Waals forces between the silicate layers result in easy absorption of hydrophilic molecules and the ability to expand perpendicular to the layers. This leads to the separation of individual clay layers in these polymers.

Most polymers, however, are hydrophobic and are not compatible with hydrophilic clays. In this case, pretreatment of either the clays or the polymers is necessary. The most popular methods for clay modification are the use of amino acids[53], organic ammonium salts[54], or tetra organic phosphonium[55] to convert the clay surface from hydrophilic to organophilic. The clays modified in this way are known as organoclays. For those polymers without any polar functional groups, such as PP, it is common to apply the techniques of grafting polar functional groups onto the polymer chains or adding grafted polymers during processing. For example, maleic anhydride grafted PP[56] has been used to produce clay/PP nanocomposites directly. The latter development also used a mixture of PP, maleic anhydride grafted PP, and organoclay[57].

Several methods have been developed to produce clay/polymer nanocomposites. Three methods were developed in the early stages of this field and have been applied widely. These are: *in situ* polymerization, solution induced intercalation [10], and melt processing [58]. *In situ* polymerization involves inserting a polymer precursor between clay layers and then expanding and dispersing the clay layers into the matrix by polymerization. The initial work in this area was carried out by the Toyota Research Group to produce clay/nylon-6 nanocomposites 26[, 27, 28, 29, 30, 59 and 60]. This method is capable of producing well-exfoliated nanocomposites and has been applied to a wide range of polymer systems. The technology is suitable for raw polymer manufacturers to produce clay/polymer nanocomposites in polymer synthetic processes and is also particularly useful for thermosetting polymers.

The solution-induced intercalation method applies solvents to swell and disperse clays into a polymer solution. This approach poses difficulties for the commercial production of nanocomposites for most engineering polymers because of the high costs of the solvents required and the phase separation of the synthesized products from those solvents. There are also health and safety concerns associated with the application of this technology. However, solution-induced intercalation is applicable to water-soluble polymers, because of the low cost of using water as a solvent and its low health and safety risks, and can be used in the commercial production of nanocomposites.

The melt processing method induces the intercalation of clays and polymers during melt. The efficiency of intercalation using this method may not be as high as that of *in situ* polymerization [61] and often the composites produced contain a partially exfoliated layered structure. However, the approach can be applied by the polymer processing industry to produce nanocomposites based on traditional polymer processing techniques, such as extrusion and injection molding. Therefore, the technology has played an important role in speeding up the progress of the commercial production of clay/polymer nanocomposites.

In addition to these three major processing methods, other fabrication techniques have been also developed. These include solid intercalation[62], covulcanization[63], and the sol-gel method[64]. Some of these methods are in the early stages of development and have not yet been widely applied.

THE CHALLENGE TO FIBER COMPOSITES

Emerging nanocomposite technology is already creating a stir in the reinforcement field. Is it reasonable to consider that clay/polymer nanocomposites could be used to replace traditional fiber reinforced composites?

In theory, the reinforcement of polymers on the nanometer scale has great advantages over traditional fiber reinforced composites. The weakness of fiber reinforced composites lies in the failure of fully utilizing the inherent properties of the materials[65]. Using carbon fibers as an example, the principle of the approach is to make carbon materials stronger by utilizing the strong covalent bonds within the aromatic sheets of the graphite structure. However, current carbon fibers only achieve 3-4% of the theoretical strength of the aromatic sheets. The lack of interconnection between the aromatic sheets in the carbon fiber structure prevents the materials from achieving high strength[65]. This is not a problem in layered-filler reinforced polymer nanocomposites. Once the layers are exfoliated in the polymer matrix, they are interconnected by the polymer so that the inherent properties of the individual layers can be fully utilized in the nanocomposite.

In reality, the mechanical properties achieved in the best clay/polymer nanocomposites are much lower than conventional fiber reinforced composites with a high fiber volume fraction. At present, the best improvement in mechanical properties has been achieved by clay/nylon-6 nanocomposites with ~4 wt% clay loading. Fig. 2 shows the tensile strength and modulus of this nanocomposite together with the original nylon-6 and nylon-6 reinforced with 48 wt% chopped glass fibers. It can be seen that the best clay/polymer nanocomposites cannot match fiber reinforced composites with a high fiber volume fraction. In order to achieve better mechanical properties, more reinforcing elements are required in clay/polymer nanocomposites. However, it may not be possible to produce a highly exfoliated nano-reinforcing structure at high clay loading levels. As long as exfoliation occurs, the surface area of the reinforcing phase will increase hundreds or thousands of times. Such a huge surface area may result in insufficient polymer molecules to wet the clay surface.

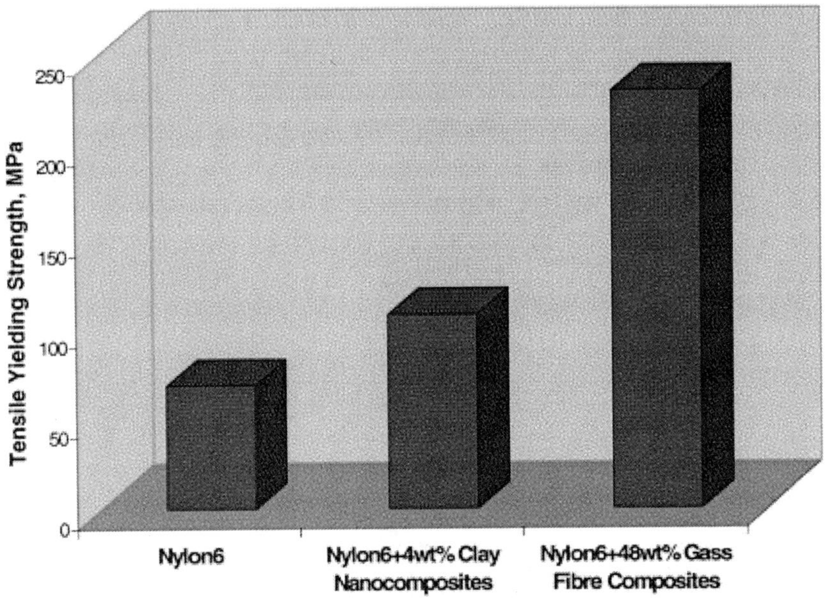

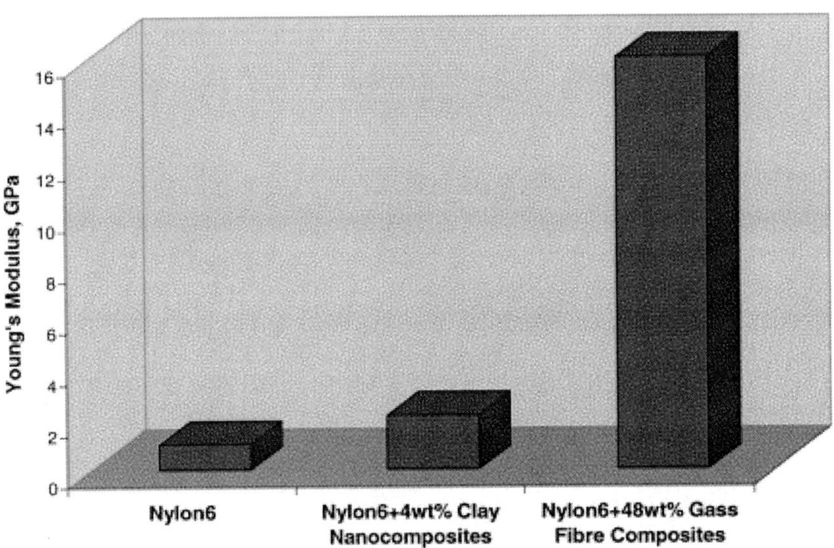

Figure 2. Comparison of the tensile strength and modulus of the best clay/nylon-6 nanocomposites and a glass fiber reinforced nylon-6 composite with 48 wt% fiber content. The properties of the original nylon-6 are also shown.

However, when clay/polymer nanocomposites and fiber reinforced composites are compared in the low filler range, nanocomposites exhibit better reinforcement than conventional fiber composites. The data, obtained by Fornes and Paul[4] for the Young's modulus of clay/nylon-6 nanocomposites and glass fiber reinforced nylon-6 composites in the filler loading range up to 10 wt%, are plotted in Fig. 3. It can be observed that the nanocomposites have a higher efficiency in improving the Young's modulus than fiber composites.

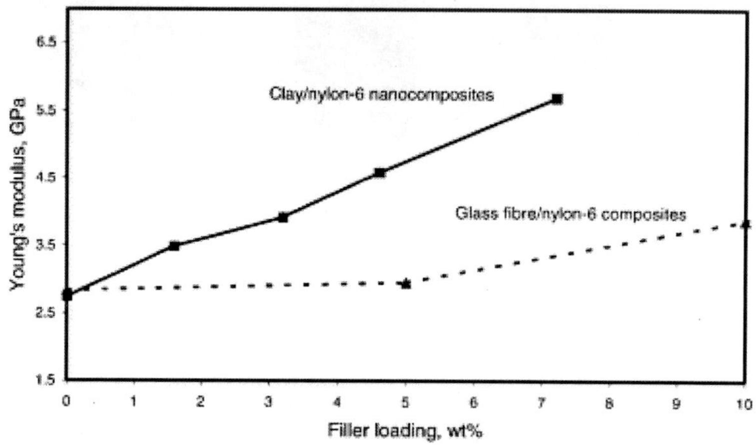

Figure 3. Comparison of the Young's modulus of clay/nylon-6 nanocomposites and glass fiber reinforced nylon-6 composites with low filler loading. The chart was plotted based on the data published by Fornes and Paul[4]

It is clear from the above comparison that clay/polymer nanocomposites have advantages over fiber reinforced composites in the low filler loading range. The market for traditional fiber composites with low fiber volume fraction could be affected by the development of clay/polymer nanocomposites. However, the progress in nanocomposites has had little impact on the high-performance fiber reinforced composite market.

DIFFICULTIES IN CLAY/POLYMER NANOCOMPOSITE DEVELOPMENT

In addition to filler loading, the major difficulties facing the development of clay/polymer nanotechnology are the lack of understanding of the mechanisms of the enhancement effect, the difficulty in application to thermosetting polymers, and the lack of thermally stable organoclays.

Although a great deal of modeling has been carried out to further the understanding of the mechanisms of enhancement of major physical and engineering properties using clay/polymer nanocomposites, we are a long way from the end of the tunnel. For example, the exact mechanical properties of individual silicate layers are not known. Hence, it is difficult to develop a mechanism of reinforcement. The structure of the char formed from the combustion of clay/polymer nanocomposites is not yet clear[66]. Without this it is not possible to develop a mechanism of fire retardancy. Extensive modeling and fundamental experimental research should be conducted in the future to remove these hurdles.

The application to thermosetting polymers is another major difficulty in clay/polymer nanocomposite development. Intercalation of clays with the precursor of a thermosetting polymer can change the functionality of the polymer. The change in functionality affects the extent of cross-linking. It is well known that the major engineering properties of thermosetting polymers are a function of the extent of cross-linking. Nevertheless, there have been reports of improvements in mechanical properties of low extent cross-linked thermosetting polymer systems such as epoxy resin with low Tg and polyurethanes. However, little progress has been made with high extent cross-linked polymer systems.

The last difficulty is directly concerned with the commercialization of clay/polymer nanotechnology, i.e. the lack of commercially available and thermally stable organoclays. The issue of the thermal stability of organoclays has now been well documented. Most commercially available organoclays are produced by the exchange of metal cations in clay galleries with organic ammonium salts. These ammonium salts are thermally unstable and can degrade at temperatures as low as 170°C[67]. Clearly, such surfactants are not suitable for most engineering plastics when applying melt processing technology to produce nanocomposites. The long-term stability of clay/polymer nanocomposites based on ammonium-salt-modified organoclays produced using other processing techniques has also become a concern. Although many thermally stable surfactants have been identified, such as phosphonium, these surfactants are too expensive for commercial use. New initiatives are moving toward direct modification of hydrophilic clays using multifunctional polymers and oligomers to produce thermally stable organoclays for clay/polymer nanocomposites.

CONCLUSIONS

Significant progress in the development of clay/polymer nanocomposites has been made over the past one and a half decades. The advantages and limitations of the technology have become clear. However, we have a long way to go before we understand the mechanisms of the enhancement of major engineering properties of polymers and can tailor the nanostructure of these composites to achieve particular engineering properties. In the low filler loading range, clay/polymer nanocomposites have the potential to replace traditional fiber reinforced composites.

REFERENCES

1. Z. Wang et al. Geophysics, 66 (2) (2001), p. 428
2. D.A. Brune, J. Bicerano Polymer, 43 (2) (2002), p. 369
3. J.-J. Luo, I.M. Daniel Compos. Sci. Technol., 63 (11) (2003), p. 1607
4. T.D. Fornes, D.R. Paul Polymer, 44 (17) (2003), p. 4993
5. O.L. Manevitch, G.C. Rutledge J. Phys. Chem. B, 108 (4) (2004), p. 1428
6. S.S. Ray, M. Okamoto Progress Polym. Sci., 28 (11) (2003), p. 1539
7. Y. Kojima et al. J. Polym. Sci., Part A: Polym. Chem., 31 (7) (1993), p. 1755
8. J.W. Gilman Appl. Clay Sci., 15 (1-2) (1999), p. 31
9. A.B. Morgan et al. J. Polym. Mater. Sci. Eng., 83 (2000), p. 57
10. K. Yano et al. J. Polym. Sci., Part A: Polym. Chem., 31 (10) (1993), p. 2493
11. C. Nah et al. Polym. Adv. Technol., 13 (9) (2002), p. 649
12. G. Gorrasi Polymer, 44 (8) (2003), p. 2271
13. R.A. Vaia Adv. Mater., 7 (1995), p. 154

14. J.C. Hutchison et al. Chem. Mater., 8 (8) (1996), p. 1597
15. Auto applications of drive commercialization of nanocomposites. Plastic Additives & Compounding(January 2002), 30
16. Cox, H., et al., Nanocomposite systems for automotive applications. Presented at 4th World Congress in Nanocomposites, EMC, San Francisco, 1-3 September 2004
17. Patterson, T., Forte™ nanocomposites - our revolutionary breakthrough. Presented at 4th World Congress in Nanocomposites, EMC, San Francisco, 1-3 September 2004
18. Goldman, A. Y., and Copsey, C. J., Multilayer barrier liner material with nanocomposites for packaging applications. Presented at 4th World Congress in Nanocomposites, EMC, San Francisco, 1-3 September 2004
19. Conway, R., Using nanotechnology to provide active and passive barriers in beverage packaging application. Presented at The Future of Nanomaterials Conference, Pira, Birmingham, 29-30 June 2004
20. Radford, J., Enhanced barrier properties in containers and films. Presented at The Future of Nanomaterials Conference, Pira, Birmingham, 29-30 June 2004
21. Carter, L. W., et al., US Patent 2,531,396 (1950)
22. D.J. Greenland J. Colloid Sci., 18 (1963), p. 647
23. A. Blumstein J. Polym. Sci., 3 (1965), p. 2665
24. B.K.G. Theng Clays Clay Miner., 30 (1982), p. 1
25. G. Lagaly Solid State Ionics, 22 (1) (1986), p. 43
26. Okada, A., et al., US Patent 4,739,007 (1988)
27. Kawasumi, M., et al., US Patent 4,810,734 (1989)
28. A. Usuki et al. J. Mater. Res., 8 (5) (1993), p. 1179
29. Y. Kojima et al. J. Mater. Res., 8 (5) (1993), p. 1185
30. Y. Kojima et al. J. Appl. Polym. Sci., 49 (7) (1993), p. 1259
31. M. Kawasumi et al. Macromolecules, 30 (20) (1997), p. 6333 e.g.
32. J.S. Bergman et al. Chem. Commun., 21 (1999), p. 2179 e.g.
33. X. Fu, S. Qutubuddin Mater. Lett., 42 (1-2) (2000), p. 12 e.g.
34. C. Wan et al. Polym. Test., 22 (4) (2003), p. 453 e.g.
35. L.W. Jang et al. J. Polym. Sci., Part B: Polym. Phys., 39 (6) (2001), p. 719 e.g.
36. M. Okamoto et al. Polymer, 41 (10) (2000), p. 3887 e.g.
37. G. Zhang et al. Mater. Lett., 57 (12) (2003), p. 1858 e.g.
38. M. Alexandre et al. Macromol. Rapid Commun., 22 (8) (2001), p. 643 e.g. View Record in Scopus
39. K.A. Carrado, L. Xu Chem. Mater., 10 (5) (1998), p. 1440 e.g.
40. X. Huang et al. Macromolecules, 33 (6) (2000), p. 2000 e.g.
41. W. Krawiec et al. J. Power Sources, 54 (2) (1995), p. 310 e.g.
42. J.M. Brown et al. Chem. Mater., 12 (11) (2000), p. 3376 e.g.

43. J.-H. Chang, K.M. Park Polym. Eng. Sci., 41 (12) (2001), p. 2226 e.g.
44. R.S. Sinha et al. Macromolecules, 35 (8) (2002), p. 3104 e.g.
45. J. Hao et al. J. Appl. Polym. Sci., 86 (3) (2002), p. 676 e.g.
46. M.H. Choi et al. Chem. Mater., 12 (10) (2000), p. 2977 e.g.
47. H.-C Lee et al. Appl. Clay Sci., 21 (5-6) (2002), p. 287 e.g.
48. S.H. Hong et al. Curr. Appl. Phys., 1 (6) (2001), p. 447 e.g.
49. Y.P. Wu et al. J. Appl. Polym. Sci., 82 (11) (2001), p. 2842 e.g.
50. H.-M. Park et al. Macromol. Mater. Eng., 287 (8) (2002), p. 553 e.g.
51. T.-K. Chen et al. Polymer, 41 (4) (2000), p. 1345 e.g.
52. R. Dhamodharan et al. J. Appl. Polym. Sci., 82 (3) (2001), p. 555 e.g.
53. A. Usuki et al. J. Mater. Res., 8 (5) (1993), p. 1174
54. K.A. Carrado Appl. Clay Sci., 17 (1-2) (2000), p. 1
55. Singh, A., and Haghighat, R., US Patent 6,057,035 (2000)
56. M. Kato et al. J. Appl. Polym. Sci., 66 (9) (1997), p. 1781
57. P. Reichert et al. Macromol. Mater. Eng., 275 (1) (2000), p. 8
58. R.A. Vaia et al. Chem. Mater., 5 (12) (1993), p. 1694
59. A. Okada et al. Mater. Res. Soc. Proc., 171 (1990), p. 45
60. Y. Kojima et al. J. Polym. Sci., Part A: Polym. Chem., 31 (4) (1993), p. 983
61. Gao, F., et al., From burning behaviour to understand the mechanisms of solution and solid intercalations in clay/PEO nanocomposites. Presented at Organic-Inorganic Hybrids II, PRA, Guildford, 28-29 May 2002
62. F. Gao et al.J. Mater. Sci. Lett., 20 (19) (2001), p. 1807
63. A. Usuki et al. Polymer, 43 (8) (2002), p. 2185
64. P. Musto et al. Polymer, 45 (5) (2004), p. 1697
65. Gao, F., e-Polymers (2002), T_004
66. Gao, F., et al., A mechanistic understanding of fire retardancy of carbon nanotube/EVA composites and their clay hybrids. Presented at the 4th World Congress in Nanocomposites, EMC, San Francisco, 1-3 September 2004
67. Gao, F., et al., The current problems with the use of reactive melt processing to produce clay/polymer nanocomposites. Presented at Organic-Inorganic Hybrids II, PRA, Guildford, 28-29 May 2002

Chapter 6

ASSESSING MECHANICAL PROPERTIES OF NATURAL FIBER REINFORCED COMPOSITES FOR ENGINEERING APPLICATION

Olusegun David Samuel[1*], Stephen Agbo[2], Timothy Adesoye Adekanye[3] [1]

[1]Department of Mechanical Engineering, Olabisi Onabanjo University, Ago-Iwoye, Nigeria

[2]Department of Mechanical Engineering, Lagos City Polytechnic, Ikeja, Nigeria

[3] Department of Agricultural Engineering, Landmark University, Omuaran, Nigeria

ABSTRACT

Mechanical properties of ukam, banana, sisal, coconut, hemp and E-glass fiber reinforced laminates were evaluated to assess the possibility of using it as new material in engineering applications. Samples were fabricated by the hand lay-up process (30:70 fibre and matrix ratio by weight) and the properties evaluated using the INSTRON material test-ing system. The mechanical properties were tested and showed that glass laminate has the maximum tensile strength of 63 MPa, bending strength of 0.5 MPa, compressive strength of 37.75 MPa and the impact strength of 17.82 J/m^2. The ukam plant fiber laminate has the maximum tensile strength of 16.25 MPa and the impact strength of 9.8J/m among the natural fibers; the sisal laminate has the maximum compressive strength of 42 MPa and maximum bending strength of 0.0036 MPa among the natural fibers. Results indicated that natural fibers are of interest for low-cost engineering applications and can compete with artificial glass fibers (E-glass fiber) when a high stiffness per unit weight is desirable. Results also indicated that future research towards significant improvements in tensile and impact strength of these types of composites should focus on the optimization of fiber strength rather than interfacial bond strength.

INTRODUCTION

Research and development of natural fibers as reinforcement for automotive sectors is a growing interest to scientists and engineers. Nowadays, natural fibers form is an interesting option for the most widely applied fiber in the composite technology. Many studies on natural studies such as keraf, bagasse, jute, ramie, hemp and oil palm [1-9]. Fiber reinforced composites with thermoplastic matrices have successfully proven their high qualities in various fields of engineering application. However, Natural fibers generally have poor mechanical properties compared with synthetic fibers but these composites were used as a source of energy to make shelters, clothes, construction of weapons [10]. High cost of synthetic fibers and health hazards of abettors fibers have really necessitated the exploration of natural fibers [11]. Consequently, natural fibers have always formed wide applications from the time they gained commercial recognition. They possess desirable properties such as biodegrability, renewability, combustibility, lower durability, excellent mechanical properties, low density and low price. Stamboulis and Baley [12] reported that this excel-lent price-performance ratio at low weight in combination with the environmentally friendly character is very important for the acceptance of natural fibers in large volume engineering markets, such as the automotive and construction industries.

Since the beginning of human existence, people have developed plant fiber composites. Brahmakumar et al. [13] reported that these composites were used as a source of energy to make shelters, clothes, construction of weapons. The use of fibers as cloth was centered largely in the country side, with higher quality textiles being available in the towns. In late medieval Germany and Italy, fibers were employed in cooked dishes, as fillers in pies and boiled soup.

A composite may be defined as a physical mixture of two or more different materials. The mixture has proper-ties which are generally better than those of any one of the materials. It is necessary to use combinations of materials to solve problems because any one material alone cannot do so at an acceptable cost or performance. These composites were produced in simple shapes and easy design structures by positioning the structural elements on top of each other to create the desired design. *Corresponding author.

Strength of glass fiber reinforced composites depends not only on the properties of the components but also on the mechanism of composite failure which is a function of how well the composite was formed [14].

Oladele et al. [15] reported that the fibre/matrix has an important role in the micromechanical behavior of composite. Lack of good adhesion with the polymeric matrix, and large moisture absorption of natural fibers adversely affect adhesion with h hydrophobic matrix material .These problems often lead to premature ageing by degrading and l oss of strength.

Natural fibre-reinforced composites have been increasingly utilized in quite widespread applications. Natural fibers are obtained from different parts of the plants, to name a few, for example jute, flax, kenaf, coconut, hemp, ukam, sisal, banana, pineapple fibers from the leaf; cot-ton and kapok from seed; coir and coconut from the fruit. For example hemp, jute, flax and sisal fibers are already used in automotive industry [16].

In polymeric composite terms, natural fiber reinforcement is a manufactured assembly of long or short bundles of natural fibers to produce a flat sheet or mat of one or more layers of fibers. These layers are held together either by mechanical interlocking of the fibers them-selves or with a binder to hold these materials together giving the assembly sufficient integrity to be handled. The un-reinforced plastics have low density, are relatively easy to process, resistant to weathering and do not require a surface finish.

The components of natural fibers are cellulose, hemi-cellulose, lignin, pectin, waxes and water soluble sub-stances. The cellulose, hemicellulose and lignin are the basic components of natural fibers, governing the physical properties of the fibers [17]. In order to fully utilize the natural fibers, understanding their physical and mechanical properties is vital. A unique characteristic of natural fibers is depended upon the variations in the characteristics and amount of these components, as well as difference in its cellular structure. Therefore, to use natural fibers to its best advantages and most effectively in automotive and industrial application, physical and mechanical properties of natural fibers must be considered. Many studies have investigated the properties of natural fibers. Numerous researchers have studied mechanical properties of varied natural fibers [18]. Traditionally, natural fibers are used and known for rope, twine, and course sacking materials; and they are biodegradable and environmentally friendly crop. All mentioned studies have assisted engineers with the design and efficient usage of the natural fibers. However, the fibers modification is required and needed to improve mechanical properties for composites product. Efficiency of the fiber-reinforced composites also depends on the manufacturing process that the ability to transfer stress from the matrix to fiber [19]. In this study, natural fiber laminated laminates were made by hand lay-up method and their mechanical properties were investigated in or-der to assess its suitability.

MATERIAL AND METHODS

Materials and Equipment

The composite materials used in the production of the specimens include: E-glass fiber (artificial fibers) ukam plant, resin, fibers, wax, release agent, gel coat and miscellaneous items. The equipment used are weighing balance, cloth, stirrers, measuring cylinder, universal testing machine.

Fiber

Treatment In this study, chemical resetting was used. The procedure involves NaOH solution treatment, water washing and drying. Natural fibers are extracted from their parent plant. The ukam, sisal and banana are extracted from the back of their stems, while hemp and coconut are extracted from their fruits. The natural fibers, after being extracted, are washed with water to remove gums. The fibers are then treated with sodium hydroxide solution and rammed. The treated fiber was allowed to dry in the sun for 3 days. After which the fibers are laid in the mold with the resin at the ratio of 30% to 70%. It was allowed to cure for about 20 days.

Laminate Manufacture

Methods In this study, samples of laminates were made by using hand lay-up method. The method used in this study was employed due to its simplicity and availability of the items. Details of the procedures taken in the production of laminates via hand lay-up method are elsewhere dis-cussed (Agbo, 2009).

Measurement

Tensile Test

The tensile tests were performed using a testing machine model 8889. The width and the thickness of the specimens were measured and recorded (360 mm by 20 mm by 5 mm). The tensile tests were carried out according to ASTM D 038-01. The tensile strengths were calculated from this test.

Bending

Test Three point bending tests were performed using a testing machine in accordance to ASTM D 790 standards. For the bending test, samples with dimensions of 300 mm × 20 mm × 5 mm were used. The bending strength test was carried out on the tonometer with its attachment fixed properly bending strengths were evaluated.

Izod Impact Test

The impact strength of notched specimen was determined by using an impact tester according to ASTM D 256-05 standards. In each case three specimens were tested to obtain average value.

RESULTS AND DISCUSSION

The test results are shown and discussed in this section. Average values of three replications of the tensile test, the bending test, the compressive test, and the impact test are tabulated in Table 1

Compressive and Tensile Strength

Figures 1 and 2 show the compressive and tensile strength of the alkalized treatment of ukam, banana, sisal, coconut, hemp and E-glass fibers. From the histograms, it is obvious that sisal laminate displayed the highest (42.0 MPa) compressive strength, followed by ukam laminate, then E-glass laminate, while the banana showed the lowest (16.75 MPa) compressive strength. Figure 2 shows the measured tensile strength of treated natural fibers. The tensile strength decreases from E-glass laminate with highest (63 MPa) tensile strength, ukam laminate (16.25 MPa), while hemp laminate showed the least (7.0 MPa) tensile strength. However other parameters that are important are assessed for good suitability.

Figure 3 shows the bending strength of alkalized natural fibers. E-glass displayed highest (0.50 MPa), next to it is sisal laminate, followed by coconut and hemp, while ukam and banana showed the least (0.0013 MPa) bending strength.

Table 1. Mechanical properties of various composite laminate

Mechanical properties	Ukam fibre Laminate	Banana fibre laminate	Sisal fibre Laminate	Coconut fibre laminate	E-glass laminate	Hemp fibre laminate
Compressive strength (MPa)	39.25	16.75	42.00	30.35	37.75	29.75
Tensile strength (MPa)	16.25	6.50	5.40	3.20	63.00	7.00
Bending strength (MPa)	0.0013	0.0013	0.0036	0.0021	0.500	0.0017
Impact strength (J/m^2)	9.89	7.47	8.36	8.36	17.82	7.41

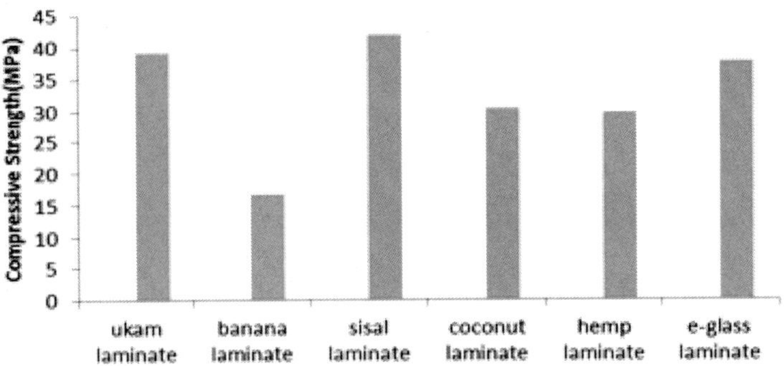

Figure 1. Compressive strength of alkalized treatment of natural fibre reinforced laminate samples.

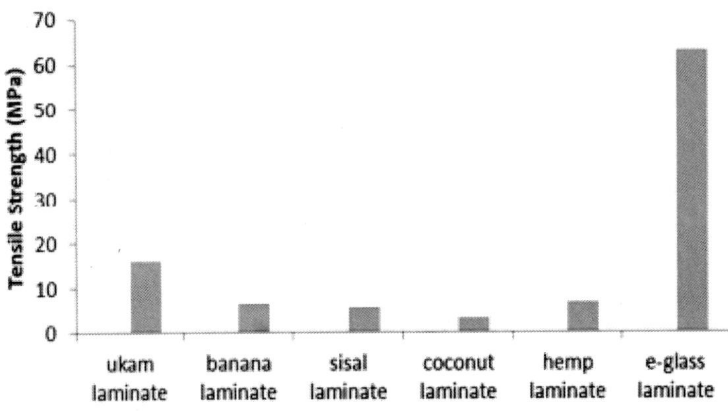

Figure 2. Tensile strength of alkanized treatment of natural fibre reinforced laminate samples.

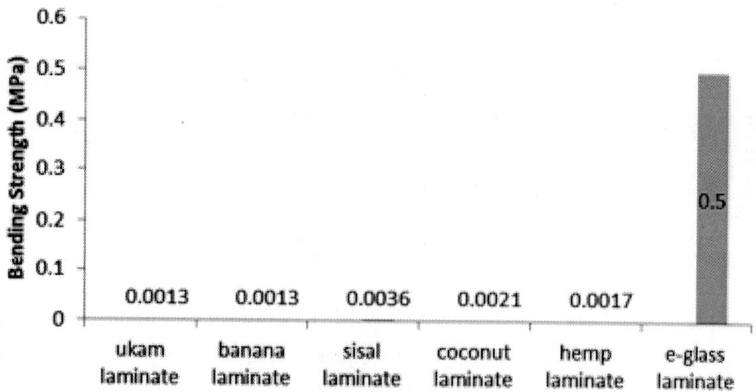

Figure 3. Bending strength of alkanized treatment of natural fibre reinforced laminate samples.

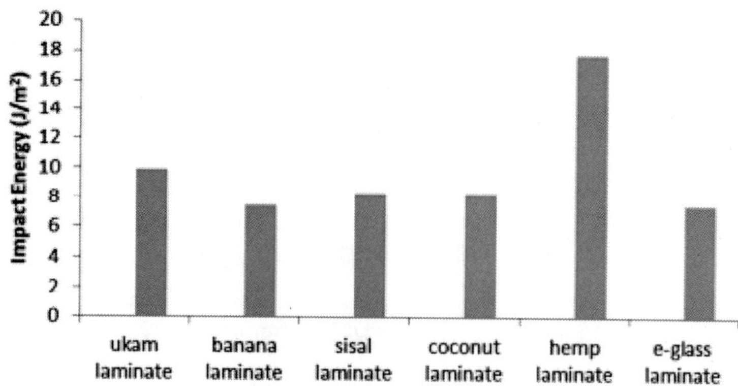

Figure 4. Impact strength of balkanized treatment of natural fibre reinforced laminate samples

Impact Strength

Figure 4 shows the izod impact strength results of rein-forced fibre laminates. The ukam laminate tested dis-played higher impact strength (9.87 J/m2) next to E-glass laminate (17.82 J/m2), while the hemp and bananana dis-played the lowest (7.477 J/m2). The alkalization treatment of fibers helps in improving the chemical bonding between the resin and fiber resulting in superior mechanical properties. It has been re-ported by several authors that mechanical properties of composites were improved by the modification of fibers [20].

CONCLUSIONS

The experimental investigation on mechanical properties of natural fiber reinforced composites leads to the following conclusions: 1) The natural fiber composite manufactured by hand lay-up process provides an opportunity of replacing existing materials with a higher strength, low cost alternative that is environmentally friendly. 2) Mechanical properties viz., Compressive strength, Bending strength, Tensile strength, and Impact strength of the ukam and sisal fiber reinforced composite material is greatly influenced by alkalization treatment. Hence, ukam and sisal fibers can be good reinforcement candi-dates for high performance polymer composites. 3) Ukam and sisal composites manufactured by hand lay-up process provide an opportunity of replacing existing materials with a higher strength, low cost alternative that is environmentally friendly.

REFERENCES

1. R. Karnani, M. Krishnan and R. Narayan, “Biofiber-Re- inforced Polypropylene Composites,” Polymer Engineer-ing and Science, Vol. 37, No. 2, 1997, pp. 476-483.
2. A. M. Mohd Edeerozey, M. A. Harizan, A. B. Azhar and M. I. Zainal Ariffin, “Chemical Modification of Kenaf Fibers,” Materials Letters, Vol. 61, No. 10, 2007, pp. 2023-2025. doi:10.1016/j.matlet.2006.08.006
3. E. Robson, “Surface Treatment of Natural Fibre,” EC/ 4316/92, 1993.
4. H. A. Sharifah and P. A. Martin, “The Effect of Alkaliza-tion and Fibre Alignment on the Mechanical and Thermal Properties of Kenaf and Hemp Bast Fibre Composites: Part 1—Polyester Resin Matrix,” Composites Science and Technology, Vol. 64, No. 9, 2004, pp. 1219-1230.
5. B. F. Yousif, K. J. Wong and N. S. M. El-Tayeb, “An Investigated on Tensile, Compression and Flexural Prop erties of Natural Fibre Reinforced Polyester Composites,” ASME International Mechanical Engineering Congress and Exposition, Seattle, 11-15 November 2007, pp. 619- 624.
6. Y. Mohd Yuhazri, P. T. Phongsakorn and H. Sihambing, “A Comparison Process between Vacuum Infusion and Hand Lay-Up Method toward Kenaf/

Polyester Compos-ite," International Journal of Basic & Applied Sciences, Vol. 10, No. 3, 2010, pp. 63-66.

7. T. Nishino, K. Hirao, M. Kotero, K. Nakamae and H. Inagaki, "Kenaf Reinforced Biodegradable Composite," Composites Science and Technology, Vol. 63, No. 9, 2003, pp. 1281-1286. doi:10.1016/S0266-3538(03)00099-X
8. H. A. Sharifah, P. A. Martin, J. C. Simon and R. P. Simon, "Modified Polyester Resins for Natural Fibre Compos-ites," Composites Science and Technology, Vol. 65, No. 3-4, 2005, pp. 525-535. doi:10.1016/j.compscitech.2004.08.005
9. P. Wamubua, J. Ivens and I. Verpoest, "Natural Fibers: Can They Replace Glass in Fibre Reinforced Plastics?" Composites Science and Technology, Vol. 63, No. 9, 2003, pp. 1259-1264. doi:10.1016/S0266-3538(03)00096-4
10. C. Benjamin and Tobias, "Fabrication and Performance of Natural Fiber-Reinforced Composite Material," 35th International SAMPLE Symposium and Exhibition, Ana- heim, 2-5 April 1990, pp. 970-978.
11. S. Agbo, "Modelling of Mechanical Properties of a Natu-ral and Synthetic Fiber-Reinforced Cashew Nut Shell Resin Composites," M.Sc. Thesis, University of Nigeria, 2009.
12. A. Stamboulis and C. Baley, "Effects of Environmental Conditions on Mechanical and Physical Properties of Flax Fibers," Composites Part A: Applied Science and Manu-facturing, Vol. 32, No. 8, 2001, pp. 1105-1115.
13. M. Brahmakumar, C. Pavithran and R. M. Pillai, "Coco-nut Fibre Reinforced Polyethylene Composites: Effect of Natural Waxy Surface Layer of the Fibre on Fibre/Matrix nterfacial Bonding and Strength of Composites," Composites Science and Technology, Vol. 65, No. 3-4, 2005, pp. 563-569.
14. M. Hautala, A. Pasila and J. Pirila, "Use of Hemp and Flax in Composite Manufacture: A Search for New Pro-duction Methods," Composite Part A: Applied Science and Manufacturing, Vol. 35, No. 1, 2004, pp. 11-16.
15. I. O. Oladele, J. A. Omotoyinbo and J. O. T. Adewara, "Investigating the Effect of Chemical Treatment on the Constituents and Tensile Properties of Sisal Fibre," Jour- nal of Minerals and Materials Characterization and Engineering, Vol. 9, No. 6, 2010, pp. 569-582.
16. M. Jacob, S. Thomas and K. T. Varughea, "Mechanical Properties of Sisal/ Oil Palm Hybrid Fibre Reinforced Natural Rubber," Composites Science and Technology, Vol. 64, No. 7-8, 2004, pp. 955-965. doi:10.1016/S0266-3538(03)00261-6
17. A. Pelet, S. Sueki and B. Mobasher, "Mechanical Proper-ties of Hybrid Fabrics in Pultruded Cement Composites," 16th European Conference of Fracture, Special Sympo-sium Measuring Monitoring and Modelling Concrete properties Alexandrroupolis, Greece, 2006.
18. H. P. S. A. Khalil and H. D. Rozman, "Rice-Husk Poly-ester Composites:

The Effect of Chemical Modification of Rice Husk on the Mechanical and Dimensional Stabil-ity Properties," Polymer Plastic and Technology Engi-neering, No. 39, 2007, pp. 757-781.

19. S. Shibata, Y. Cao and I. Fukumoto, "Press Forming of Short Natural Fiber-Reinforced Biodegradable Resin: Ef-fects of Fiber Volume and Length on Flexural Proper-ties," Polymer Testing, Vol. 24, No. 8, 2005, pp. 1005- 1011. doi:10.1016/j.polymertesting.2005.07.012

20. M. Idricula, S. K. Malhota, K. Joseph and S. Thomas, "Dynamic Mechanical Analysis of Randomly Oriented Intimately Mixed Short Banana/Sisal Hybrid Fibre Rein-forced Polyesters Composites," Composites Science and Technology, Vol. 65, No. 7-8, 2005, pp. 1077-1087. doi:10.1016/

Chapter 7

INFLUENCE OF THE COMPOSITE SURFACE STRUCTURE ON THE PEEL STRENGTH OF METALLIZED CARBON FIBRE-REINFORCED EPOXY

E. Njuhovic[a], A. Witta, M. Kempf[a], F. Wolff-Fabris[a, 1], S. Glöde[b], V. Altstädt[a]

[a] University of Bayreuth, Department of Polymer Engineering, Universitaetsstraße 30, 95447, Bayreuth, Germany

[b] Lüberg Elektronik GmbH & Co. Rothfischer KG, Hans-Striegl-Straße 3, 92637 Weiden, Germany

ABSTRACT

In this work, the effect of mechanical pre-treatment on the surface structure of carbon fibre-reinforced epoxy composites and on its peel strength of electroless/electroplated copper was investigated. Sandblasting with Al_2O_3 was used to pre-treat the composite surface. The parameters investigated were blasting time (3 s, 6 s and 9 s) and nozzle distance to substrate (300 mm and 500 mm). A two-step metallization process was used for depositing copper coatings on the pre-treated composite surface. First, an eletroless plating process was used to deposit a thin layer on the surface. Second, an electroplating process was used to reinforce the thickness of the coating. Increased blasting intensity leads to a significant increase in surface roughness, which promotes mechanical anchoring effects of the coating. Scanning electron microscopy images and contact angle measurements confirm the results of the surface roughness. The adhesion of sandblasted composites, characterized by measuring the peel strength, is 10 times higher compared to untreated specimens. In addition to the mechanical anchoring mechanism the exposure of carbon fibres on the surface due to the blasting process promoted a stronger bonding to copper, due to the higher, electrical conductivity of the fibres in comparison to the matrix.

INTRODUCTION

The usage of carbon fibre-reinforced polymers (CFRP) within the automotive and aerospace industry is continuously growing in the last decades to replace traditional materials and achieve weight reduction. Currently, this is not only the case for structural parts, but also for storage systems for cryogenic liquid hydrogen, which is gaining attention from both academic and industrial communities.

Liquid hydrogen as an energy carrier is of special interest because of the much higher gravimetric energy density compared to gaseous and solid stored hydrogen as well as compared to conventional fuel systems. Prototypes and technology demonstrators using liquid hydrogen can be found in automobile and aerospace industries [1]. Traditionally made of stainless steel, the cryogenic storage systems can alternatively be manufactured with CFRP leading to a weight reduction of approximately 60% [1].

However, standard epoxy composites are not sufficiently tight for the storage of liquid hydrogen because of the higher permeation and outgassing rates compared to stainless steel. A metal coating on the surface of the polymer composite is therefore required as permeation barrier in order to fulfill the requirements [1].

Suitable coating processes of CFRP are vacuum-metallization (e.g. PVD or CVD), indirect metallization (e.g. hot foil stamping) and plating processes (e.g. electroless/electrolytic plating) [1], [2] and [3].

Hot foil stamping is a suitable and economically viable method for relatively simple 2D geometries [2]. However, it cannot be employed for the manufacture of cryogenic storage systems. In the case of these complex-shaped 3D parts manufactured with CFRP, plating process is the most suitable coating process mainly because of faster deposition rates, higher ductility of the coatings and lower process temperatures compared to PVD or CVD processes [4].

Regardless which of the above-mentioned processes is selected to coat the CFRP with a metallic layer for permeation barrier purposes, it is generally very difficult to create consistently high adhesive strength levels between the composite and the coating materials [5]. This is due to the much lower polarity of the polymer surface in comparison to the coating material [6]. As consequence of the weak adhesion, the coating can detach from the CFRP surface leading to a significant permeability increase. This process is further accelerated due to the fact that storage systems are subjected to dynamic loadings [7]. One reason for this arises due to the difference of the outer temperature (room temperature, e.g. 23 °C) and inner temperature (− 253 °C, liquid hydrogen). Furthermore, such storage systems are subjected during their lifecycle to a number of predictable and unpredictable mechanical loadings, especially considering that these systems are employed in the transport sector. For instance, considering the promising use of cryogenic hydrogen as energy carrier for satellites, the storage system and the CFRP/coating interface must remain intact during and after the rocket launch.

To increase the adhesion of the polymer substrate with the coating layer, surfaces are often treated in a way to: a) increase the surface roughness for mechanical adhesion, or b) modify the surface energy to increase the wettability and adsorption [2] and [8].

In both cases the surface is modified by pre-treatment processes, which can be generally classified as mechanical, chemical or electrical pre-treatments. Mechanical processes are grinding and sandblasting whereas etching and wet-chemical surface modification are typical examples for chemical processes. Electrical pre-treatment processes include atmospheric and low-pressure plasma treatment [9].

Regarding thermoplastics, examples for chemical pre-treatment can be found in case of acrylonitrile butadiene styrene (ABS), the most widely electroless plated plastic. In this case chromic acid has two effects on the surface, which results in an improved adhesion. It increases the surface energy and wettability by oxidizing the surface and it dissolves the polybutadiene nodes in ABS, which increases the surface roughness and significantly improves the mechanical adhesion [10]. In case of polyetherimide etching with permanganate, the imide ring of the molecule is opened and allows the copper ions to be incorporated into the system, which results in a high adhesion of the cooper coating to the polymer substrate [11].

Examples for electrical pre-treatment are found in case of plasma treatment of polycarbonate surfaces for palladium chemisorption prior to electroless deposition. Charbonnier et al. showed that after plasma treatment a high efficiency in grafting chemical functions could be achieved [12]. Direct palladium chemisorption onto nitrogenated groups is highlighted. However no influence on the adhesion was presented.

Chemical pre-treatment of epoxy resins is very difficult due to the narrow processing window and the high chemical resistance of this thermoset to most etching media. This leads to difficulties achieving a structured surface, as either too long times or too aggressive media will lead to its destruction [13]. Effect of alkaline etching on the surface roughness of a fibre-reinforced epoxy composite has been studied by Roizard et al. [14]. It has been shown that a small change in the topography could be achieved but the effect of the adhesion of a metal coating was not investigated. Kirmann et al. [15] studied the effects of the alkaline permanganate etching of epoxy on the peel adhesion of electrolessly plated copper on a fibre-reinforced epoxy composite. In this study the adhesion could be controlled by chemical etching with alkaline permanganate. An extra epoxy layer was applied to the composite surface to avoid fibre damage during etching.

Electrical pre-treatment of epoxy-based composites can also be found in the literature, for instance in case of plasma surface treatment of carbon fibre-reinforced epoxy composites [16], [17] and [18]. Zaldivar et al. [16] studied the effect of atmospheric plasma treatment on the chemistry, morphology and resultant bonding behavior. The bonding strength after plasma treatment could be increased as much as approximately 50%. In a further study, Zaldivar et al. [17] continued the investigation of how plasma treatment process parameters affect the surface chemistry and the bonding behavior. The changes in the surface chemistry after the plasma treatment could be correlated with the adhesive bond strength. These studies looked at epoxy bonding but did not investigate metallized surfaces.

Sandblasting and its parameters as mechanical pre-treatment have been mainly investigated on metal substrates [19] and [20]. It has been shown that adhesion of a coating is strongly dependent on the surface roughness of the metal substrate, which can be regulated by blasting parameters. Generally speaking, higher blasting intensity (e.g. higher blasting pressure, lower distance, higher times) leads to a higher surface roughness [21].

For epoxy composites the influence of blasting angle on the adhesion of metal coatings was studied by Menningen et al. [5]. Advantages of this pre-treatment include higher adhesion strengths in comparison to grinding and easier processing in comparison to chemical etching [2] and [21]. Fracture mechanics was applied to the adhesion and a change from 30° to 90° blasting angle led to an increase in energy release rate of approximately 40%. But in this study no quantitative analysis of the surface structure was presented. In a further study Menningen et al. [22] investigated the effect of micro roughening on the adhesion strength of a nickel coating on a CFRP surface. It has been shown that an increase in blasting pressure leads to higher surface roughness but not necessarily to higher adhesion strength. However the influence of further blasting parameters, such as blasting time and distance, on the surface structure and adhesion strength is surprisingly still not investigated.

There is therefore a lack of knowledge in the literature in the field of copper-plated carbon fibre-reinforced epoxy composites, which we cover with this manuscript. This study focuses on the effect of the surface structure, generated with a mechanical pre-treatment method (sandblasting), on the peel strength of copper electroless-/electroplated fibre-reinforced epoxy composites. The topography and the wettability of the surface of the composite are heavily dependent on the selected pre-treatment process and its parameters [2]. This study presents a correlation between the surface properties of the substrates and the peel strength of the metallized material as well as the parameters of the pre-treatment process.

EXPERIMENTAL

Substrate Material

In this study CFRP material consisting of carbon fibres (non-woven 0°/90° biaxial NCF HS Carbon from WELA) with an areal weight of 300 g/m^2 and a toughened epoxy resin as matrix (XU3508/XB3486 from Huntsman) were used. The CFRP laminates were manufactured by VARTM-process in a 1-part machine setup with a two-sided hard mould. The application of release agent Loctite Frekote 770-NC was done thoroughly on the mould surfaces as mould preparation before injection. The laminate thickness of 2 mm corresponds to a fibre volume content of approximately 50%. The laminates were cured at 100 °C for 5 h according to the resin manufacturer's datasheet.

Surface Pre-treatment

The CFRP surface must be pre-treated prior to metallization of the material. The method investigated in this study is sandblasting with aluminum oxide and 200 – 300 µm grit size and a mohs hardness of 10. The parameters investigated are blasting time (3 s, 6 s and 9 s) and nozzle distance to substrate (300 mm and 500 mm). The depth of abrasion is dependent on the blasting time whereas the nozzle distance influences the blasting medium velocity and thus the kinetic energy of a blasting particle. The sandblasting machine ST 1200 ID-Z-SB with a die diameter of 10 mm is used to perform the tests. Constant parameters are blasting pressure of 2 bars and a blasting angle of 90°. All plates including the reference laminate were cleaned using an ultrasonic bath with equal parts of ethanol and water for 30 min at 25 °C prior to the coating process.

Mechanical Properties of the Untreated and Pre-treated Composite

The mechanical properties of the untreated and sandblasted composites were investigated under quasi-static 3-point bending using a universal testing machine Zwick Z2.5. For the sandblasted composites flexural properties of specimens exposed to the highest blasting intensity were investigated. The test was carried out according to EN ISO 14125 using a rectangular bar horizontally positioned on two supports. The specimens were subjected to a vertical force applied midway between the supports at a velocity of 2 mm/min. The specimens were prepared according to class IV in the standard with the dimensions 100 mm × 15 mm. For each of the two series five measurements were performed in order to emphasize the repeatability of the results.

Coating Process

The CFRP substrates were coated by the electroless/electrolytical plating process. Direct electrolytical plating of CFRP is impossible due to the electrical insulation of the polymer matrix. On account of this, a thin adherent conductive layer was chemically deposited on the CFRP surface. For this chemical deposition, the surface is made electrically conductive with a one-step activator to leave as many palladium ions on the surface as possible. Therefore the substrate was dipped into an aqueous solution consisting of a stabilized Pd–Sn colloid. Palladium needs to be protected in order to prevent agglomeration and drop out [23]. In an accelerator bath the enclosed Pd ions are broken free to leave palladium on the surface. At molecular level, the single palladium atoms are not homogeneously dispersed on the surface but they create clusters of molecular size. However they are packed enough in order to provide a homogeneous Cu layer. After this activating process, a 1 µm thick copper coating was deposited electrolessly on the surface and finally electrolytically plated with the same coating material. The final coating thickness was at least 40 µm. A rigorous surface preparation procedure was employed in this study. The electroless-/electroplating process is shown below including further cleaning steps which are as well applied to all plates:

1. Acidic Cleaner Circuposit™ 3323A for 5 min at 50 °C
2. Rinsing in deionized water for 1 min at RT
3. Low concentrated etching sulphuric etching for 1.5 min at 35 °C
4. Rinsing in deionized water for 1 min at RT
5. Pre-Dip Circuposit™ 3340 for 0.5 min at RT
6. Activator (solution of colloidal palladium–tin) for 4 min at

40 °C

7. Rinsing in deionized water for 1 min at RT
8. Accelerator (aqueous solution of fluoroboric acid) for 6 min at RT
9. nRinsing in deionized water for 1 min at RT
10. Electroless plating of Cu (aqueous solution of copper sulphate, ethylenediaminetetraacetic acid, sodium hydroxide, formaldehyde and sodium cyanide) for 20 min at 45 °C
11. Rinsing in deionized water for 1 min at RT
12. Pickling for 0.5 min at RT
13. Electroplating of Cu (aqueous solution of copper sulphate, sulphuric acid and sodium chloride) for 20 min at 1.8 A/dm^2
14. Rinsing in deionized water for 1 min at RT
15. Drying for 13 min at 65 °C

Surface Structure

Surface Roughness

The roughness measurements were carried out with a Universal Surface Tester 100 from Innowep GmbH.

A 60° steel cone with a radius of curvature of 30 µm was used as tip to measure the surface profile and roughness of the pre-treated samples. A constant tip force of 1 mN and a tip speed of 0.1 mm/min were set to ensure reproducible roughness measurements according to DIN EN ISO 4287 and ASTM D 7127 - 05. The surface of the samples were measured by 10 lines with a parallel distance of 2 mm and a measuring length of 20 mm in order to obtain representative information about the roughness.

Microscopy

Surface investigations of the CFRP substrates were carried out by light and electron microscopy.

An optical microscope, Keyence VHX 100, was used, to look at the deposits on the interfacial side of the coatings and on the substrates after the mechanical testing of the adhesion strength.

A scanning electron microscope (SEM), Jeol JSM-IC 848, was used to inspect the topography of the untreated and pre-treated CFRP surfaces. The samples were gold sputtered prior to the SEM investigation.

Contact Angle and Surface Energy

In measuring the contact angle the surface tension of pre-treated samples and consequently the degree of wettability are determined. A drop of fluid with a defined volume and known surface tension is applied on the sample surface. The contact angle θ is then measured in the three-phase system solid (S), liquid (L) and gaseous (G) [2].

Contact angles of test liquids were recorded using a goniometer. The liquids tested were distilled water, and diiodomethane. Five droplets of each liquid, 2 µl in volume, were measured on each samples surface, in order to measure the contact angle. For each sample, the mean and standard deviation were calculated. Using the Owens, Wendt, Rabel, and Kaeble method, surface energy values were calculated as follows [24]:

$$\frac{(1+\cos\theta)\cdot\gamma_L}{\sqrt{\gamma_L^d}} = \sqrt{\gamma_s^p}\cdot\sqrt{\frac{\gamma_L^p}{\gamma_L^d}} + \sqrt{\gamma_s^d}$$

θ is the static contact angle of the liquid on the surface of the polymer, γ_L is the surface energy of liquid and is taken from the literature [25] and γ_s is the surface energy of the polymer. Superscripts d and p represent the dispersive and polar components of the surface energy accordingly. It should be noted that the total surface energy of a liquid or solid is equal to the sum of the dispersive and polar components.

Peel Strength

The peel test was carried out according to ASTM B 533-85 using a universal testing machine, Zwick Z2.5. A 25 mm wide metal stripe was cut out of the substrate, using a paper knife, torn off at one end and peeled off at a velocity of 25 mm/min. The force was recorded as a function of the measuring path by the software. To calculate the peel strength the mean of the recorded force was used and divided by the width of the peeled stripes. The peeling orientation was parallel to the fibres. The testing standard used does not provide any specific orientation for the specimens. Some unpublished investigations of our group comparing pulling orientation 0° and 45° did not show any difference in peel strength.

RESULTS AND DISCUSSION

In order to assure that all composite plates had a sufficient curing and the same conditions for further processing steps, DSC measurements according to ISO 14322 were performed after production of the laminates, surface pre-treatment and coating process. After the initial curing cycle the system had a degree of cure above 98% and no further change was observed after the subsequent steps.

Surface Structure After Pre-treatment

The pre-treatment process influences the surface structure of the composite and consequently the topography of the substrates. Especially the surface roughness is an important aspect in correlating the adhesion strength to the topography of the substrates.

Due to the manufacturing process, the CFRP laminates exhibited a closed epoxy matrix layer at the surfaces. Therefore, it is possible to pre-treat the samples within the epoxy matrix layer on the one hand and on the other hand to remove the outer layer while carbon fibres become exposed to some extent. Damage of the fibres can be expected and can lead to lower mechanical properties. In order to investigate this aspect, the mechanical properties of the untreated and sandblasted composites were also investigated under quasi-static 3-point bending. Fig. 1 shows the results of the bending properties of an untreated specimen and after sandblasting for 9 s and 300 mm distance. A minor performance loss of approximately 5% was observed after sandblasting. The fatigue properties of the mentioned materials (including low temperature cycling) are currently under investigation and will be the topic of a subsequent manuscript.

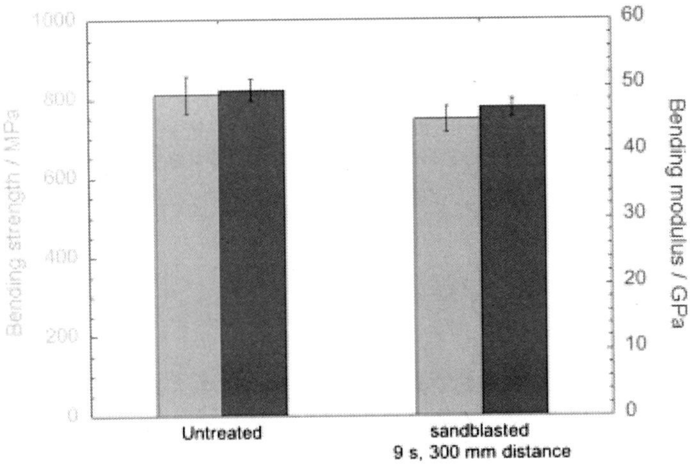

Figure.1. 3 point bending test of an untreated and sandblasted CFRP.

The variation of the blasting parameters time and distance results in a significant increase in surface roughness. Table 1 shows the mean roughness index, the roughness depth and the contact angle of sand blasted carbon fibre-reinforced epoxy composites as a function of the blasting parameters. As expected, it is clearly visible that the surface roughness is increased at higher blasting times and at a reduced distance. The mean roughness index *Ra* is approx. 18 times higher in case of the longest blasting time and smallest distance than the untreated reference sample. On the other hand the increase at a distance of 500 mm compared to 300 mm did not show a major effect on the roughness as the particles experience only a minor kinetic energy loss when traveling such short distances (200 mm) in air.

Table 1. Surface roughness and contact angle of a blasted carbon fibre-reinforced epoxy composite as a function of different blasting parameters

Parameters		R_a	R_z	Contact angle
Distance (mm)	Time (s)	(μm)	(μm)	(deg)
Reference		0.30 ± 0.07	2.94 ± 1.05	106 ± 5
300	3	1.56 ± 0.26	17.68 ± 3.41	141 ± 12
	6	2.96 ± 0.27	26.32 ± 1.88	139 ± 11
	9	5.33 ± 0.61	37.81 ± 3.76	135 ± 7
500	3	1.16 ± 0.29	12.74 ± 3.50	111 ± 5
	6	2.69 ± 0.71	23.86 ± 4.12	139 ± 11
	9	3.84 ± 0.42	33.28 ± 4.15	140 ± 4

The effect of mechanical pre-treatment on the surface structure of carbon fibre-reinforced epoxy composite substrates is also seen when comparing the contact angles of untreated and blasted samples. The untreated sample shows low wettability referring to the contact angle of approximately 102°. After blasting with Al_2O_3 the contact angle increases and does not further significantly change at longer blasting times in case of a blasting distance of 300 mm. A significant change is merely noticeable between 3 s and 6 s at a distance of 500 mm. The increase in the contact angle can be attributed to the capillary depression effect. With increased surface roughness the adhesion properties decrease because the fluid is not pulled into the cavities but remains on top of the embossment of the surface.

Sandblasting of the CFRP samples generates a non-uniform surface structure characterized by dimples and furrows. The difference of an untreated and pre-treated sample surface is clearly visible in Fig. 2, Fig. 3, Fig. 4 and Fig. 5. At longer blasting times more epoxy resin of the outer layer is removed so that the carbon fibres are also exposed and damaged (Fig. 5). According to Table 1, the depth of the abrasion and the wastage rate seems to be lower at a distance of 500 mm compared to a distance of 300 mm. The influence of the blasting time and distance shown in the SEM images is reflected in the surface roughness measurements. Comparing the untreated specimen to the material blasted for 3 s a major change on surface roughness can be observed. However, longer times, namely 6 and 9 s only led to minor changes in surface roughness.

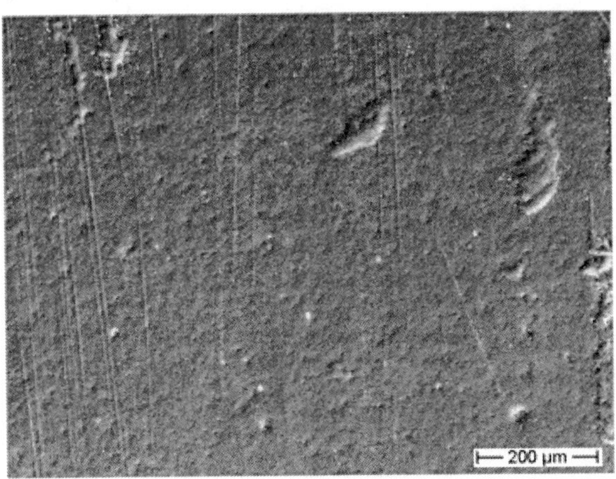

Figure 2.SEM image of an untreated CFRP substrate.

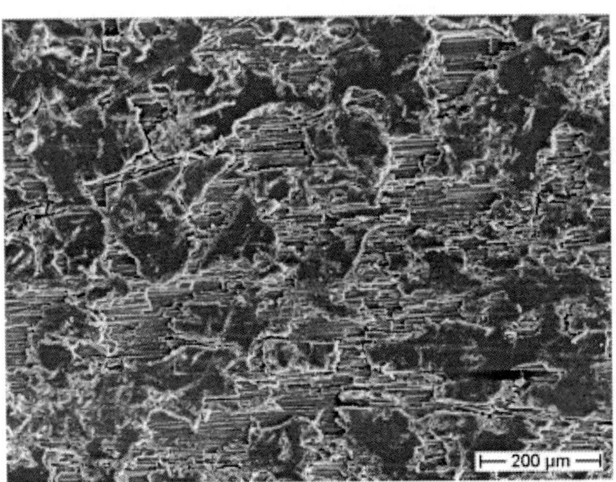

Figure 3.SEM image of an Al_2O_3 blasted CFRP substrate (time 3 s, distance 500 mm).

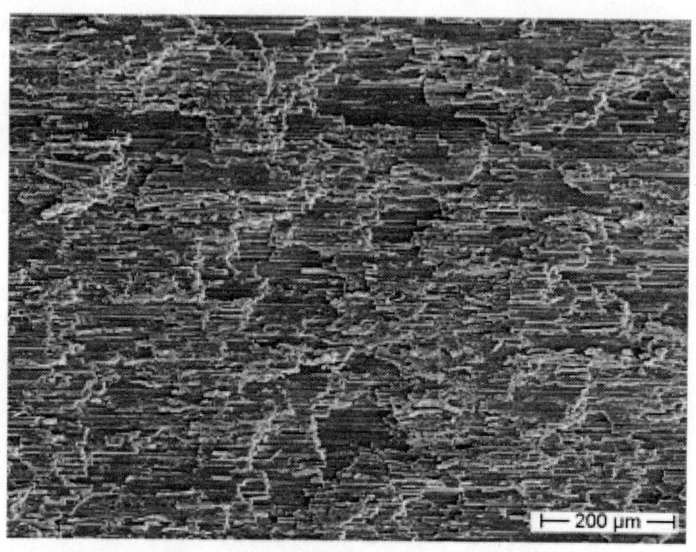

Figure 4. SEM image of an Al_2O_3 blasted CFRP substrate (time 6 s, distance 500 mm).

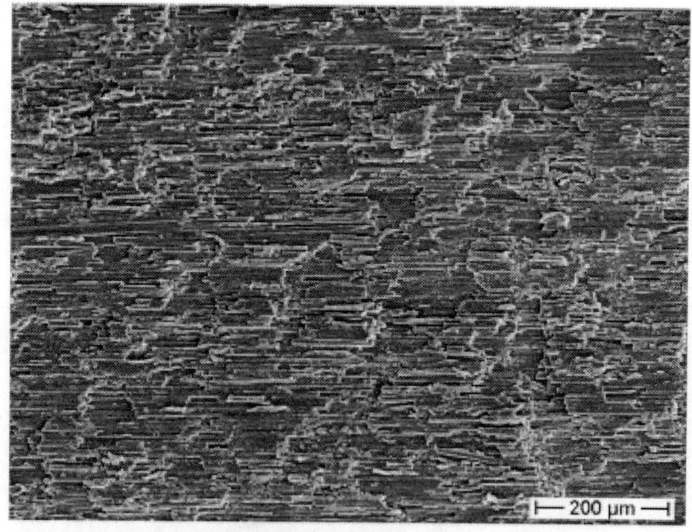

Figure 5. SEM image of an Al_2O_3 blasted CFRP substrate (time 9 s, distance 500 mm).

CFRP/Metal Layer Adhesion

The force required to separate a metallic coating from its polymer substrate is determined by the interaction of several factors: the components and quality of the polymer moulding compound, the moulding process, the process used to prepare the substrate for electroplating, and the thickness and mechanical properties of the metallic coating. If only one of the parameters is changed at time, peel strength can be used to quantify the influence of this parameter on the adhesion strength. Fig. 6 shows the influence of the blasting time and distance on the peel strength of copper-coated CFRP substrates. As expected, an increase of the roughness leads to interlocking effects at the surface and consequently higher peel strength. It can clearly be seen that blasting time plays a role on the peel strength, whilst no significant difference is observed between the two blasting distances investigated here. The peel strength of blasted samples, even for 3 s blasting time, is approximately 10 times higher than that of untreated samples. An increase of the blasting time, which leads to a rougher surface, does not lead to a major further increase of the peel strength though. This behavior is seen in Fig. 7. One explanation for that is the fact that the copper coating has a minimum thickness of only 40 µm, and by reaching a substrate roughness around this value interfacial weak points are formed. This can be seen in Fig. 8 and Fig. 9 where cross-section images of sandblasted and copper-coated CFRP substrates are shown. The increase in surface roughness changes the appearance of the coating, as the metallization process is kept constant. Comparing these two figures one can see that in the case of the sandblasted metallized substrate with a higher blasting intensity the copper layer is in direct contact with the carbon fibres, while in the sandblasted composite with a lower blasting intensity the copper is mostly at the top of the resin-rich zones. As there was no significant difference in the images between blasting time 6 and 9 s, only one image is presented here. The exposure of carbon fibres due to sandblasting was previously observed and described (Fig. 3, Fig. 4 and Fig. 5). This aspect should also be taken into account when considering the increase of the peel strength observed in this study. Due to their higher electrical conductivity, carbon fibres exposed at the surface promote the deposition of electroplated copper on the surface on a more readily and stronger way than the non-conductive epoxy layer. This leads to a local stronger bonding force between the layer and the copper, resulting in higher peel strength. This behavior can be observed looking at the fracture surfaces of the peel-strength-specimens, Fig. 10, Fig. 11 and Fig. 12.

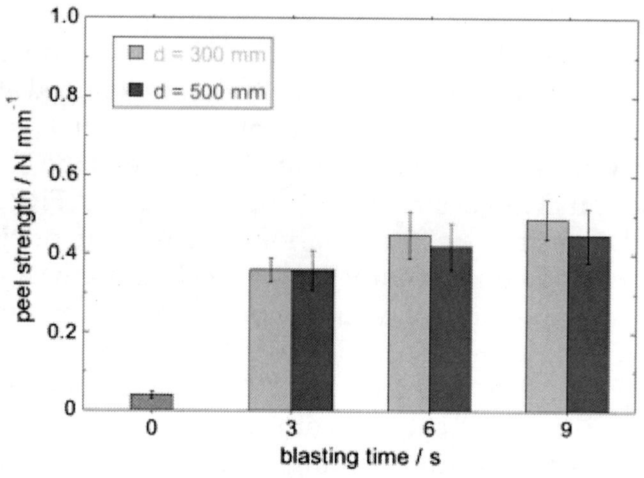

Figure 6. Peel strength as a function of blasting time and nozzle distance d.

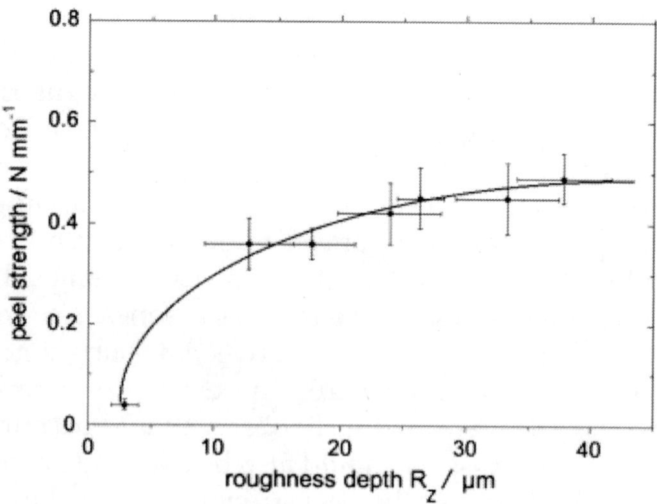

Figure 7. Peel strength as a function of roughness depth R_Z.

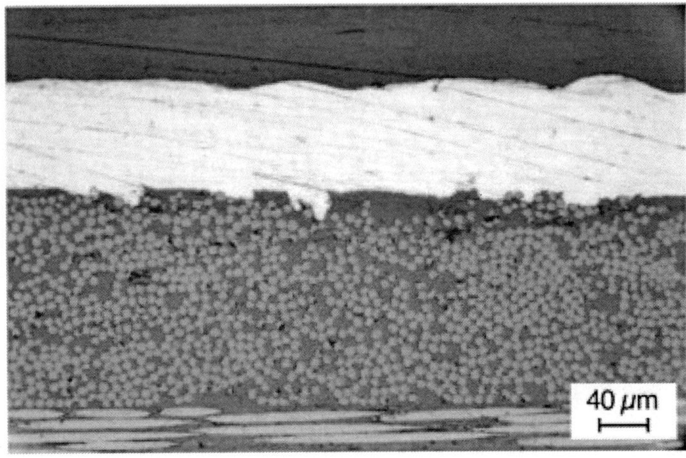

Figure 8. Cross section of copper-coated CFRP substrate after sandblasting with Al_2O_3 (500 mm and 3 s).

Figure 9. Cross section of copper-coated CFRP substrate after sandblasting with Al_2O_3 (500 mm and 6 s).

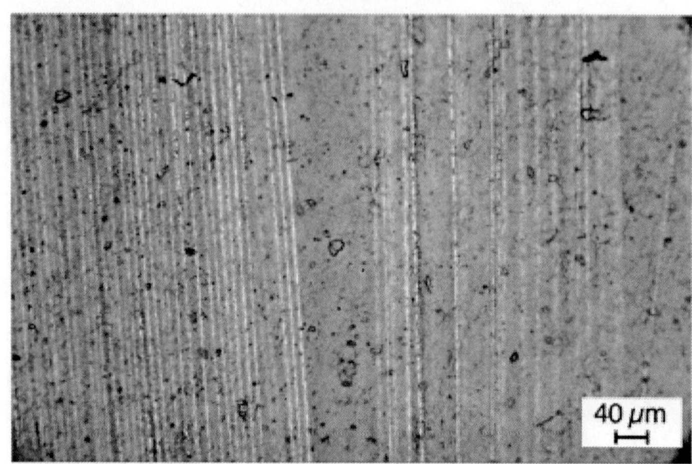

Figure 10. Fracture surface of an untreated CFRP substrate after peel test under optical microscopy.

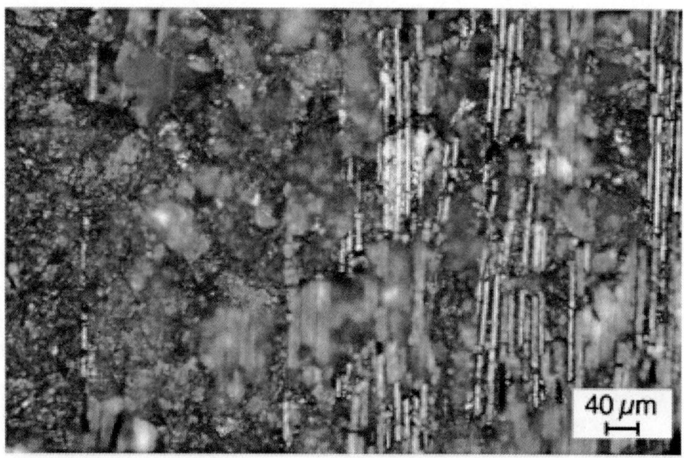

Figure 11. Fracture surface of an Al_2O_3 blasted CFRP substrate after peel test under optical microscopy (blasting time 9 s, blasting distance 500 mm). Amplification magnitude: 20 ×.

Figure 12. Fracture surface of an Al_2O_3 blasted CFRP substrate after peel test under optical microscopy (blasting time 9 s, blasting distance 500 mm). Amplification magnitude: 50 ×.

Fig. 10 shows the surface of an untreated CFRP surface, which is featureless, indicating a very low adhesion between substrate and copper. On the other hand, copper residues can be observed at the fracture surfaces of the blasted specimens. Moreover, these residues are located directly over the carbon fibres, which were exposed after the blasting treatment. This demonstrates the stronger affinity between the high electrical conductive carbon fibres and copper in comparison to the epoxy/copper interface. The surface of the peeled metal coating was observed using light microscopy. The back side of copper layers removed from the substrate surface after peel test are seen in Fig. 13 and Fig. 14. Fig. 13 shows the back side of a Cu layer of an untreated CFRP surface whereas in Fig. 14 a Cu layer back side of a sandblasted CFRP surface (9 s, 500 mm) is visible. There is no presence of fibres or resin debris on the peeled metal. The specimens with higher adhesion present simply a metal coating, which copies the surface structure of the substrate (Fig. 14). In such cases there is evidence of cohesive failure of the coating (as cooper debris are present on the substrate) but there is no evidence of cohesive failure of the laminate.

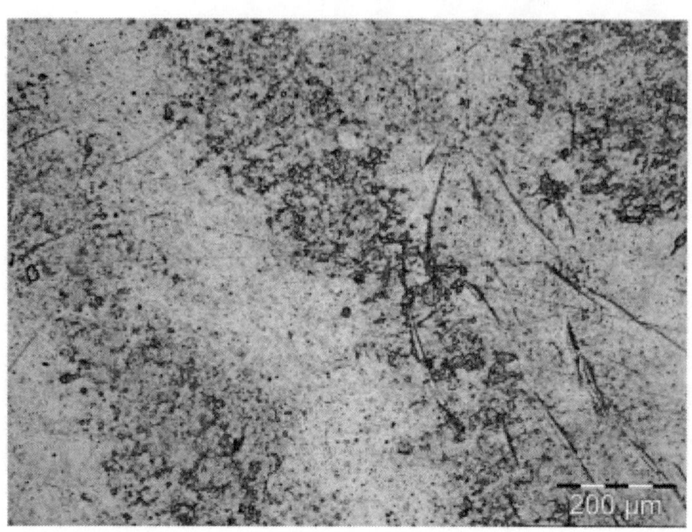

Figure 13. Fracture surface of the back side of a Cu layer of an untreated CFRP substrate after peel test under optical microscopy.

Figure 14. Fracture surface of the back side of a Cu layer of an Al_2O_3 blasted CFRP substrate after peel test under optical microscopy (blasting time 9 s, blasting distance 500 mm).

CONCLUSIONS

In this work, the effect of mechanical pre-treatment on the surface structure of carbon fibre-reinforced epoxy composites and on its peel strength of electroless/electrodeposited copper was investigated. The pre-treatment method employed here was sandblasting with Al_2O_3, where blasting time and distance were changed. An impressive 10-fold increase of the peel strength was obtained in this study.

On one hand, this increase was due to the higher surface roughness of the substrate after pre-treatment. Simultaneously, the contact angle also increased due to the capillary depression effect. The roughness was mainly influenced by the blasting time and to a lower extent by the blasting distance.

On the other hand, the peel strength of the blasted specimens also increased due to the exposure of carbon fibres at the surface. Due to the higher electrical conductivity of carbon fibres in comparison to epoxy, the electroplating process was locally facilitated and a stronger adhesion of the copper coating to the exposed carbon fibres could be reached. To our best knowledge this is the first time that such behavior is reported in the literature, and leads to new possibilities and approaches for the development of future modern metallized CFRP.

ACKNOWLEDGMENTS

The authors wish to thank the Zentrales Innovationsprogramm Mittelstand (ZIM), which is the project executing organization in the frame of the project KF2116705MF9

REFERENCES

1. D. Schultheiss, Permeation Barrier for Lightweight Liquid Hydrogen Tanks, Univerity of Augsburg, 2007.
2. R. Suchentrunk, Kunststoff-Metallisierung, Handbuch für Theorie und Praxis, Saulgau/Württ, Eugen-Leuze, 1991.
3. K. Holmberg, A. Matthews, Coatings Tribology – Properties, Techniques and Applications in Surface Engineering, Elsevier Tribology Series, 28, Elsevier Science B.V., The Netherleands, 1994.

4. J. Fessmann, D. Mann, G. Kampschulte, F. Leyendecker, T. Bolch, K. Mertz, J. Surf. Coat. Technol. 54 (55) (1992) 599.
5. M. Menningen, H. Weiss, J. Surf. Coat. Technol. 76 (77) (1995) 835.
6. H. Hofmann, J. Spindler, Verfahren der Obe fl ächentechnik, 2004, (Leipzig).
7. V.L. Morris, Proc 14th Int SAMPE Symp Company Report: Stuctural Composites, Pomona, California, U. S. A., 1989, p. 1867.
8. S. Karakoca, H Y ı lmaz, Wiley InterScience DOI: http://dx.doi.org/10.1002/jbm.b.31477 , 2009.
9. P. Gupta, G. Tenhundfeld, E.O. Daigle, D. Ryabkov, J. Surf. Coat. Technol. 201 (2007) 8746.
10. A. Garcia, J. Appl. Mater. Interfaces 2 (2010) 1177.
11. G. Porta, D. Foust, M. Burrell, B. Karas, J. Polym. Eng. Sci. 32 (15) (1992).
12. M. Charbonnier, M. Alami, M. Romand, J. Electrochem. Soc. 143 (1996) 472.
13. J.A. Koutsky, J.S. Mijovic, J. Polym. Plastics Technol. Eng. 9 (2) (1977) 139.
14. X. Roizard, M. Wery, J. Kirmann, J. Compos. Struct. 56 (2002) 223.
15. J. Kirmann, X. Roizard, J. Pagetti, J. Halut, J. Adhes. Sci. Technol. 12 (4) (1998) 383.
16. R.J. Zaldivar, J.P. Nokes, G.L. Steckel, H.I. Kim, B.A. Morgan, J. Compos. Mater. 44 (2010) 137.
17. R.J. Zaldivar, H.I. Kim, G.L. Steckel, J.P. Nokes, J. Compos. Mater. 44 (2010) 1435.
18. H. Li, H. Liang, F. He, Y. Huang, Y. Wan, J. Surf. Coat. Technol. 203 (2009) 1317.
19. K. Chung, B. Hsu, T. Berry, T. Hsieh, J. Oral Rehabil. 28 (2001) 418.
20. B. Lim, S. Heo, Y. Lee, C. Kim, J. Biomed. Mater. 64B (2003) 38.
21. A. Momber, Blast Cleaning Technology, Springer, Berlin, 2008.
22. M. Menningen, H. Weiss, U. Fischer, J. Surf. Coat. Technol. 71 (1995) 208.
23. G. Herrmann, Handbuch der Leiterplattentechnik, Band 3, Chapter 2 Saulgau/Württ, Eugen-Leuze, 1993. [24] D.H. Kaelble, J. Adhes. 2 (1970) 66.
24. H.J. Busscher, J. Arends, J. Colloid Interface Sci. 81 (1981) 75

Chapter 8

IMPACT BEHAVIOR OF GLASS FIBERS REINFORCED COMPOSITE LAMINATES AT DIFFERENT TEMPERATURES

Amal A.M. Badawy *

Materials Engineering Department, Faculty of Engineering, Zagazig University, Zagazig, 44519, Egypt

ABSTRACT

The impact behavior of glass fibers reinforced polyester (GFRP) was experimentally investigated using notched Izod impact test specimen. The experimental program was carried out on unidirectional laminate of GFRP in directions 0°, 45° and 90° in addition to cross-ply laminate $(0/90/\bar{0})_s$(0/90/0¯)s. The effect of fiber volume fraction, V_f% (16%, 23.2% and 34.9%) was considered. The impact specimens were tested after exposure to temperatures of −10 °C, 20 °C, 50 °C and 80 °C for exposure time of 1 h and 3 h. Test results showed that the effect of exposure temperature and fiber volume fraction on impact strength of GFRP composite depends on the parameter controlling the mode of failure, i.e. matrix or fiber. The failure characteristic changed from fiber pull-out to fiber breakage with increasing the exposure temperature.

INTRODUCTION

The response of composite materials under dynamic loading has received much attention recently. Application of the materials can be found both in the transportation industry and in military structures. These materials are being used increasingly due to their superior strength, light weight and adaptable design, where the ability to tailor their mechanical properties by the choice of fiber, fiber direction and matrix. However, in dynamic situations their behavior is very complex due to a large number of factors that govern their response. These factors include impact velocity, specimen geometry, impactor size, clamping mode, matrix properties and reinforcement geometry [1]. In contrast to metals which absorb energy in elastic–plastic deformation, fiber-reinforced polymer (FRP) exhibit a variety of fracture mechanisms including matrix deformation and micro-cracking, interfacial debonding, lamina splitting, delamination, fiber breakage and fiber pull-out [2]. The effect of fiber volume fraction on the impact performance of glass fiber reinforced polymer (GFRP) was studied by various investigations with contradictory results. It was found that, the increase in the fiber content introduced increasing in impact resistance [3], [4] and [5]. However, Khalid [6] found an opposite trend. He found that the increase in the fiber content leads to a reduction in the impact resistance.

The study of impact response of laminated composites subjected to environmental conditions other than ambient is more realistic. The investigated temperatures are express to the ambient temperature in which the composite materials are used. The previous investigations studied the effect of temperature ranging from −50 to 120 on the impact behavior of composite material. The effect of exposure temperature on the impact performance of GFRP was studied by few investigations [7], [8], [9] and [10]. It was found that, the increase in temperature leads to a decrease in the impact resistance. On the contrary, another study [6] found inverse behavior, where an increase in the impact resistance was found with increasing temperature.

Therefore, the influence of temperature on the impact behavior of composite laminate still under focus and need more study. In the present study, the impact specimens were exposed to temperatures of −10 °C, 20 °C, 50 °C and 80 °C for exposure time of 1 h and 3 h before testing. These advanced laminates are susceptible to impact damage either during assembly or in service. In particular, the internal delamination being the major mode of damage, generated often due to the relatively low interlaminar shear strengths, is not visible on the impact surface and is likely to grow under subsequent loading which can lead to a significant loss in structural stiffness, strength and lead to catastrophic failure [11], [12], [13], [14], [15] and [16]. Therefore, it is of vital importance to have a better understanding of the impact characteristics, energy absorption and induced damage of the laminated composite. Topics on this issue are still open for investigation. Therefore, the objective of this work is to study the influence of the fiber volume fraction, temperature and exposure time on the impact strength and failure modes of composite laminate.

EXPERIMENTAL WORK

Materials and Specimens

The materials used in this work were E-glass fibers in roving form as a reinforcement and polyester resin (Resipole 9588 ST) as a matrix. The methyl ethyl ketone peroxide (1% by weight of resin) was used as catalyst and Cobalt naphthenate (0.2% by weight of resin) as accelerator. The laminated composites were made in form of plates consist of five layers by hand lay-up technique (ISO 1268). The mould was hardening at room temperature for 24 h, then the plate removed from the mould and matured for 21 days before machining the specimen. The specimens were carefully cut to the required dimensions of the impact test according to ASTM D256. The notched Izod impact specimen dimensions were 63.5 mm × 12.7 mm with thickness of 4 ± 0.2 mm having V notch of 45° with a depth of 2 mm prepared by milling machine.

The experimental program was carried out on GFRP unidirectional laminates in directions 0°, 45° and 90° in addition to cross-ply laminate $(0/90/\bar{0})_s$(0/90/0¯)s to study the impact behavior with the variation of fiber volume fractions, V_f%, (16%, 23.2% and 34.9%) and exposure temperature (−10 °C, 20 °C, 50 °C and 80 °C) at exposure time of 1 h and 3 h. A brief description of the parameters investigated in the present study is illustrated in Table 1. Five specimens for each test condition were prepared. This leads to the fabrication of 265 test specimens.

Table 1. Parameters investigated in the present study

Type of material	Fiber direction	V_f%				Temperature (°C)				Exp. time (h)		No. Spec.
		0	16	23.2	34.9	−10	20	50	80	1	3	
Polyester		☼	–	–	–	☼	☼	☼	☼	☼	☼	40
Unidirectional	0°	–	☼	☼	☼	☼	☼	☼	–	–	☼	45
Laminate	45°	–	☼	☼	☼	☼	☼	☼	–	–	☼	45
	90°	–	☼	☼	☼		☼		–	–	☼	15
Cross-ply laminate	$(0/90/\bar{0})_s$	–	☼	☼	☼	☼	☼	☼	☼	☼	☼	120

Test Procedures

The 6709 (Avery-Denison limited) machine was used to conduct the impact test. A hammer and a digital attachment for recording the energy absorption were used to perform the tests. The impact speed was fixed at 3.46 m/s. The Frazer of refrigerator was used for cooling the specimens while an oven was used for heating the other specimens. An infrared thermometer (FLUKE 61) was used to check that the specimen temperature is maintained at the desired temperature. Exposure time was 1 h and 3 h. Table 2 shows the test condition and measurement gauge for the tested specimens. Computer Light Microscope, USB 2.0 digital microscope, of magnification up to 250X was used to assess the type and extent of damage.

Table 2. The condition and measurement gauge for the tested specimens

Temperature (°C)	Test condition	Measurement gauge
−10	Refrigerator	Infrared thermometer (FLUKE 61)
20	Oven	Temperature indicator
50	Oven	Temperature indicator
80	Oven	Temperature indicator

RESULTS AND DISCUSSION

Results of this study include the impact strength of unidirectional and cross-ply laminates in addition to failure mechanism and microscopic examination of composite specimens.

The Impact Energy of Unidirectional Laminate

Fig. 1 shows the variation of impact strength with fiber volume fractions for unidirectional composite laminates at different fiber orientation after exposure to temperature of 20 °C for 3 h. It is observed that, the impact strength increases with increasing V_f% for 0° unidirectional laminated specimens. However, insignificant effect of fiber volume fractions on impact strength was observed for laminates with fibers orientation of 45° and 90°. Fig. 2 illustrates the modes of failure for unidirectional composite laminated specimens with $V_f = 34.9\%$ at different fiber orientation. It is clear that, in the case of 0° unidirectional composite laminate, the presence of fibers arrest the crack propagation in the test specimen and the final failure was due to delamination between layers in the compression zone, see Fig. 2a. Therefore, fiber volume fractions plays a crucial rule in the impact strength of 0° unidirectional composite laminate. On the other hand, the cracks propagate in the direction of fibers in the case of 45° and 90° unidirectional laminated composites result in fracture in these directions, see Fig. 2b and c. Therefore, there is no significant effect of fiber volume fractions in theses fiber directions. In the case of 45° unidirectional laminated composites both fibers and matrix control the failure by different mechanism, while in the case of 90° unidirectional laminated composites the whole behavior is controlled by the matrix with no contribution to the fibers. Also, the length of crack path in the case of 45° unidirectional laminated is longer than that of 90° unidirectional laminated. Thus the impact strength of 45° unidirectional laminated composites is higher than that of 90° unidirectional laminated composites.

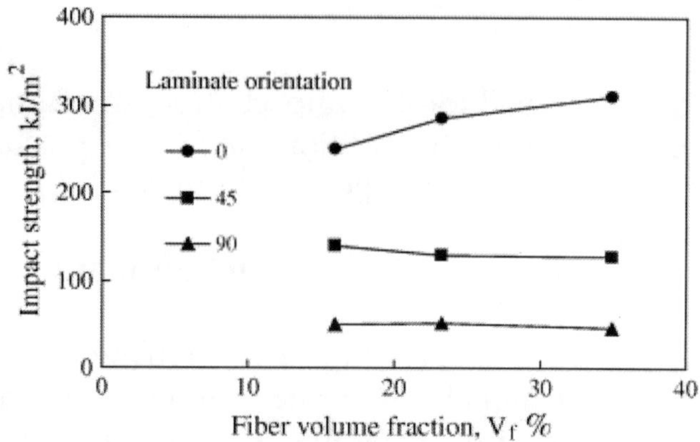

Figure 1.Effect of fiber volume fraction on impact strength for unidirectional laminates at 20 °C.

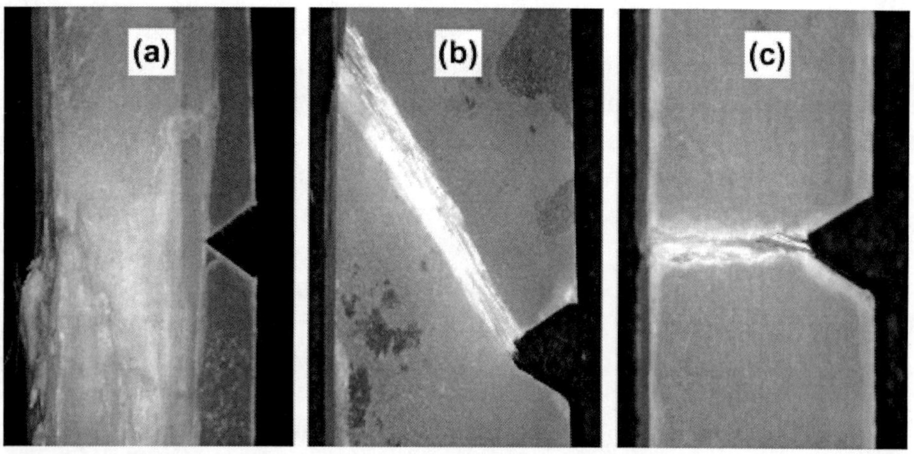

Figure 2.Failure modes of (a) 0° unidirectional laminate, (b) 45° unidirectional laminate and (c) 90° unidirectional laminate, at $V_f = 34.9\%$.

The effect of exposure temperature on impact strength of 0° unidirectional composite with different fiber volume fractions is shown in Fig. 3. The Figure illustrates that, the impact strength decreases with increasing the exposure temperature for GFRP specimens, while it increases for the polyester specimens. The variation of impact strength with exposure temperature for 45° unidirectional composite laminate specimens having different V_f% is shown in Fig. 4. As expected, the behavior of specimens with this fiber orientation is similar to the behavior of polyester (matrix) in Fig. 3. This is because the matrix controlling the behavior of laminate in this fiber orientation. The increase in temperature leads to tougher matrix (softening) consequentially more energy is required to fracture the test specimen. The figure also illustrates that the increase in V_f% leads to a decrease in the impact strength especially at low temperature. Thus, it can be concluded that the effect of exposure temperature and fiber volume fraction on impact strength of unidirectional GFRP composite depends on the parameter controlling the mode of failure, i.e. matrix or fiber.

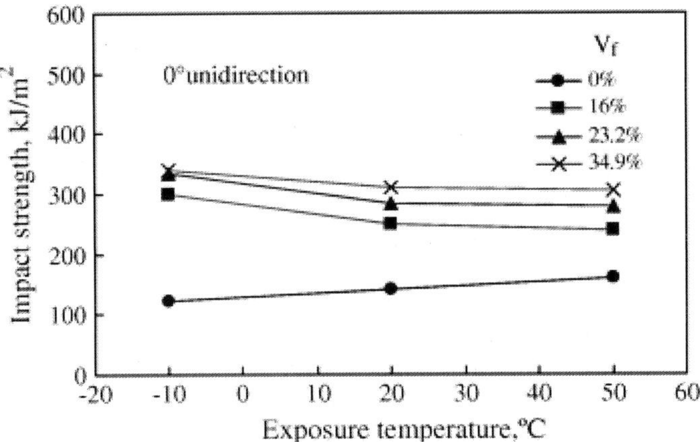

Figure 3. The variation of impact strength with exposure temperature for 0° unidirectional composite laminate.

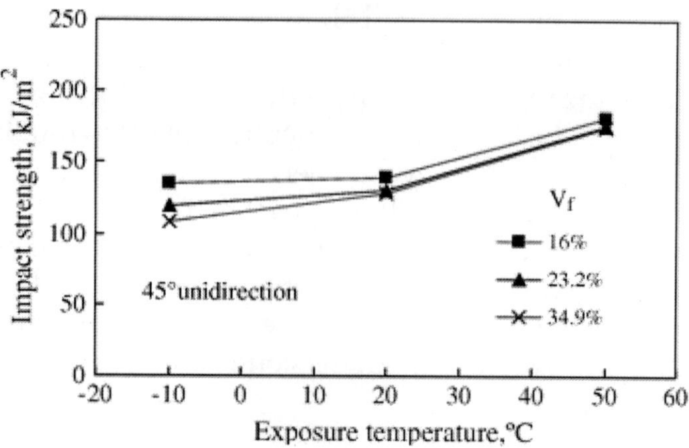

Figure 4. The variation of impact strength with exposure temperature for 45° unidirectional composite laminate.

The Impact Energy of Cross-ply Laminate

Fig. 5 shows the effect of exposure temperature for 3 h on impact strength for cross-ply laminate with different V_f%. The figure illustrates that, the impact strength decreases with increasing the exposure temperature for cross-ply GFRP specimens. The decrease in the impact strength is insignificant for low V_f%, i.e. V_f = 16%, up to exposure temperature of 80 °C. Similar trend is observed for the higher percentages of fiber volume fraction up to 50 °C. Above 50 °C, a noticeable reduction in the impact strength is recorded. The figure also shows that, the increase in V_f% leads to an increase in the impact strength of cross-ply GFRP specimens. It is clear from the impact strength results of unidirectional GFRP laminates that, 0° unidirectional composite is directly proportional to fiber volume fraction and inversely proportional to exposure temperature (fibers control laminate behavior). On the other hand, an opposite trend is observed for 90° unidirectional composites and polyester (matrix control laminate behavior). This means that the behavior of cross-ply specimens under impact loads is a combination between the behavior of 0° and 90° unidirectional composites in addition to the effect of delamination between layers.

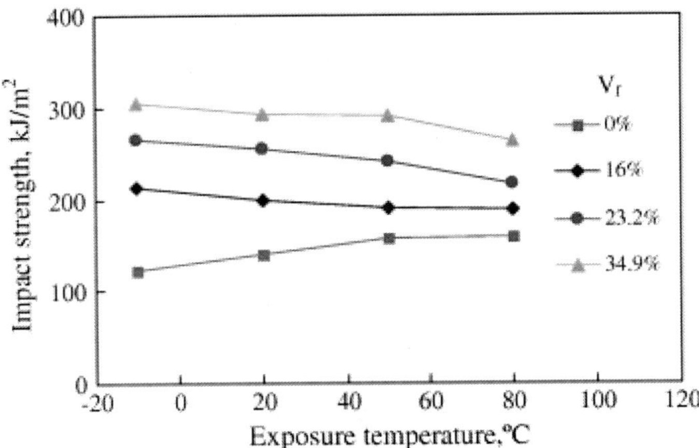

Figure 5. Effect of exposure temperature on impact strength for cross-ply laminate.

Fig. 6 shows the damage area of cross-ply specimens for different V_f% exposed to temperatures of (a) −10 °C and (b) +80 °C for 3 h. The behavior of these materials can be understood by observing the white zone near the fracture surface. It is clear that more damage, as indicated by the white zone in the images, are induced at lower temperature than at higher temperature. A possible reason for this is that, the difference between the impact strength of 0° and 90° unidirectional composites at a low temperature is much higher than that at higher temperature, i.e. more delamination at lower temperature, as shown in Fig. 6a. The reduction in white zone correlates well with create less damage and absorb less energy for specimens impacted at high temperature, as shown in Fig. 6b. The figure also indicate that increasing fiber volume fraction make the fracture difficult. The notch impact strength is generally reflects the energy required to propagate an existing notch in the specimen. The role of fibers on the propagation of a crack through a matrix is to increase the volume in which energy dissipation can take place. The presence of fibers also increases the number of potential energy absorbing mechanisms in the system.

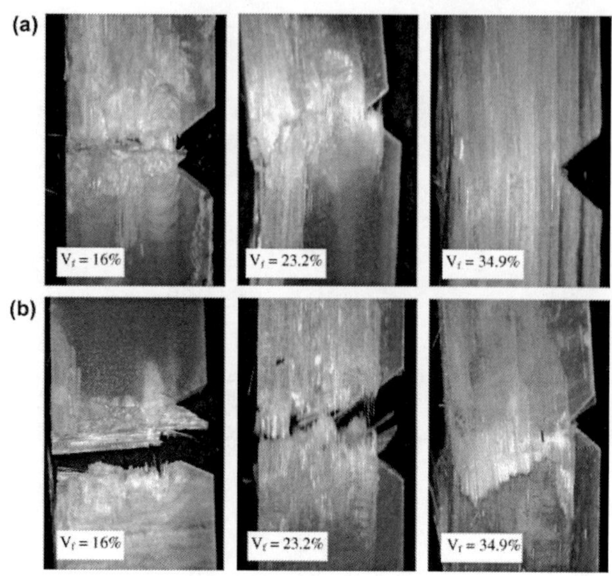

Figure 6. Impacted specimens for different V_f% exposed to temperatures of (a) −10 °C and (b) +80 °C.

This analysis can further validated by using a light microscope inspection to the fracture surface of specimens. Fig. 7 illustrates the fracture surface for cross-ply specimen with V_f = 23.2% at two different exposure temperatures of (a) −10 °C and (b) +80 °C. At low temperature (−10 °C), low bond strength between matrix and fibers is expected. This makes the fibers that fractured away from the crack interface will be pulled out from the matrix at different sections, which may involve more energy dissipation. On the other hand, increasing exposed temperature (+80 °C) leads to softening the matrix and thus enhancing the bond between fibers and matrix. This bond enhancement results in fibers breakage at approximately one section. This explanation is observed clearly in Fig. 7. In general, the failure characteristic changed from fiber pull-out to fiber breakage with increasing the exposure temperature.

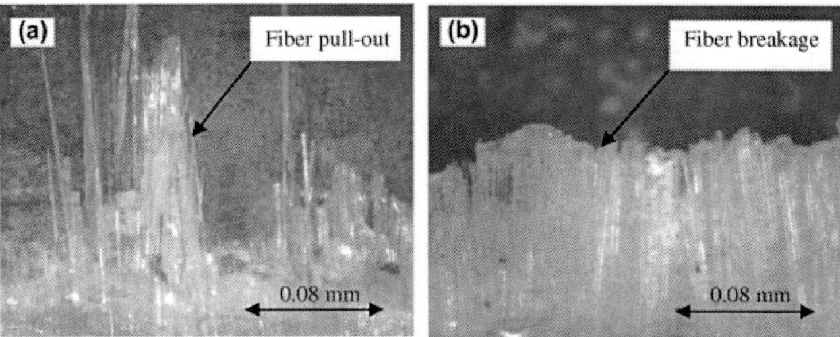

Figure 7. Fracture surface for specimens with V_f = 23.2% after exposure to temperatures of (a) −10 °C and (b) +80 °C.

The effect of exposure time on the impact strength for cross-ply laminate with different fiber volume fractions at different exposure temperatures is shown in Fig. 8. It is clear that, the impact strength of polyester specimens exposed to 3 h is higher than those exposed to 1 h for all exposure temperatures. On the other hand, the figure demonstrates a marginal effect of exposure time on the impact strength of cross-ply laminate with different fiber volume fractions at different exposure temperatures.

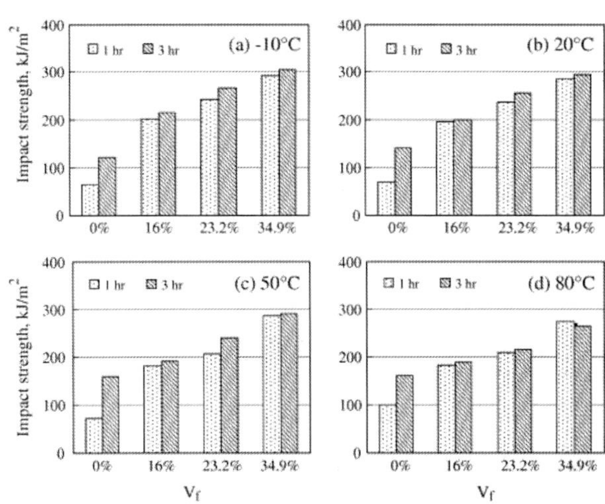

Figure 8. Effect of exposure time on impact strength for cross-ply laminate with different V_f% at different exposure temperature (a) −10 °C, (b) 20 °C, (c) 50 °C and (d) 80 °C.

Comparison Between Unidirectional and Cross-ply Laminates

Comparison between the impact strength of 0° unidirectional and cross-ply laminated composites is shown in Fig. 9. The figure illustrates that the 0° unidirectional laminated composite present higher impact strength than the cross-ply laminate within the whole temperature range and for all fiber volume fractions. This behavior may be explained as follows: using the Izod impact test, all the fibers in the unidirectional laminate specimens were perpendicular to the direction of the applied load, while only a portion of fibers were perpendicular to the direction of the applied load in the cross-ply laminate specimens. An opposite trend was found by Ibekwe et al. [9] using drop weight impact test.

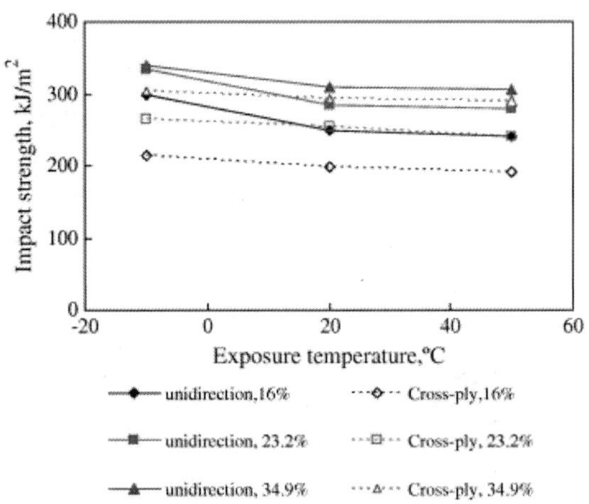

Figure 9.The variation of impact strength with exposure temperature for 0° unidirectional and cross-ply laminates.

CONCLUSIONS

From the present experimental results, the following conclusions can be drawn:

- Fiber orientation and fiber volume fraction are important microstructure parameters in the impact characterization of GFRP composite, where the impact strength increased with increasing $V_f\%$ for 0° unidirectional specimens, while insignificant effect of fiber volume fractions on impact strength was observed for 45° and 90° unidirectional composites.

- Increasing fiber volume fraction increased the impact strength for cross-ply laminated composites.
- The impact strength decreased with increasing the exposure temperature for both 0° unidirectional and cross-ply laminated composites, while increased for the polyester specimens.
- The effect of exposure temperature and fiber volume fraction on impact strength of GFRP composite depends on the parameter controlling the mode of failure, i.e. matrix or fiber.
- The impact strength of polyester specimens exposed to 3 h was higher than that exposed to 1 h for all exposure temperatures. On the other hand, a marginal effect of exposure time on the impact strength of cross-ply laminate was recorded for all temperature range.
- The 0° unidirectional laminated composite presented higher impact strength than the cross-ply laminate within the whole temperature range and for all fiber volume fractions.
- More impact damaged area was induced in specimens impacted at lower temperatures than those at higher temperatures. In general, the failure characteristic changed from fiber pull-out to fiber breakage with increasing the exposure temperature.

REFERENCES

1. Richardson MOW, Wisheart MJ. Review of low-velocity impact properties of composite materials. Composites 1996;27:1123–31.
2. Mittal RK, Jafri MS. Influence of fiber content and impactor parameter on transvers impact response of uniaxially reinforced composite plates. Composites 1995;26:877–86.
3. Thomason JL. The influence of fibre length, diameter and concentration on the impact performance of long glass-fibre reinforced polyamide. Composites 2009;40:114–24.
4. Thomason JL, Vlug MA. Influence of fibre length and concen- tration on the properties of glass fibre-reinforced polypropylene: 4. Impact properties. Composites 1997;28:277–88.
5. Cheon SS, Lee DG. Impact characteristics of glass fiber compos- ites with

respect to fiber volume fraction. J Compos Mater 2001;35:27–56.

6. Khalid AA. The effect of testing temperature and volume fraction on impact energy of composites. Mater Des 2006;27:499–506.
7. Alcock B, Cabrera NO, Barkoula NM, Wang Z, Peijs T. The effect of temperature and strain rate on the impact performance of recyclable all-polypropylene composites. Composites 2008;39:537–47.
8. Hirai Y, Hamada H, Kim JK. Impact response of woven glass- fabric composites-II. Effect of temperature. Compos Sci Technol 1998;58:119–28.
9. Ibekwe SI, Mensah PF, Li G, Pang SS, Stubblefield MA. Impact and post impact response of laminated beams at low tempera- tures. Compos Struct 2007;79:12–7.
10. Khojin AS, Bashirzadeh R, Mahinfalah M, Jazar RN. The role of temperature on impact properties of Kevlar/fiberglass composite laminates. Composites 2006;37:593–602.
11. Yang FJ, Cantwell WJ. Impact damage initiation in composite materials. Compos Sci Technol 2010;70:336–42.
12. Zaretsky E, deBotton G, Perl M. The response of a glass fibers reinforced epoxy composite to an impact loading. Int J Solids Struct 2004;41:569–84.
13. Naik NK, Meduri S. Polymer-matrix composites subjected to low-velocity impact: effect of laminate configuration. Compos Sci Technol 2001;61:1429–36.
14. Aktas M, Atas C, Icten BM, Karakuzu R. An experimental investigation of the impact response of composite laminates. Compos Struct 2009;87:307–13.
15. Feraboli P, Kedward K. A new composite structure impact performance assessment program. Compos Sci Technol 2006;66:1336–47.
16. Karakuzua R, Erbil E, Aktas M. Impact characterization of glass/ epoxy composite plates: an experimental and numerical study. Composites 2010;41:388–95.

Chapter 9

RESEARCH ON TRIBOLOGICAL BEHAVIOR OF PEEK AND GLASS FIBER REINFORCED PEEK COMPOSITE

E.Z.Li, W.L.Guo, H.D.Wang, B.S.Xu,X.T.Liu

Academy of Armored Forces Engineering, Dujiakan 21, Fengtai District, Beijing 100072, China

ABSTRACT

The tribological behaviors of pure polyetheretherketone (PEEK) and PEEK composites reinforced by 30wt% short glass fibers (GF) were comparatively evaluated on a ball-on-disc configuration at room temperature. The effects of applied load and sliding time on the friction coefficient and wear loss of the GF/PEEK were examined. The mechanical property, morphology and thermal performance of the composite were studied. The results indicated that the friction coefficient and wear loss of the composite increased gradually and tended to be a stable state as the increase of applied load and sliding time. The GF/PEEK has an excellent wear resistance, compared with PEEK. The SEM and EDS indicated that the short glass-fibers were extruded from the composite rather than pulverized into the composite. Compared with that of pure PEEK, the thermal decomposition temperature of GF/PEEK composite had an increase of 75□. The tensile strength and flexural strength of the composite were increased by 64% and 66%, respectively.

INTRODUCTION

The applications of polymers which are replacing traditional materials are increasing polymers such as polyetheretherketone (PEEK) and polytetrafluoroethylene (PTFE) are important engineering materials in recent years. PEEK is a popular matrix material for high performance composites due to its high mechanical strength and elastic modules, high melting temperature, chemical inertness, high toughness, easy processing and wear resistance.

PEEK has also many applications in engineering and medicine because of its high strength and high melting point relative to other polymers, as well as its resistance to chemical and biological action[1,2]. Vast number of investigations related to the tribological behavior of PEEK and its composite has been reported that the PEEK reinforced with some fibers has a beneficial effect on its strength and tribological properties [3, 4].The addition of short fibers that enhance the thermal conductivity are often of great advantage, especially if effects of temperature enhancement in the contact area are to be avoided in order to prevent an increase in the specific wear rate. Since glass fibers have low strength, high flexural modulus and low expansion rate, they are the most common fiber reinforcements of thermoplastics to reduce the expansion rate and increase the flexure of PEEK [5, 6].

M. Sumer and H. Ulna have studied the tribological performance of pure PEEK and 30wt% fiber glass reinforced PEEK under dry sliding and water lubricated conditions. The results showed that the friction coefficient and specific wear rates for pure PEEK and GF/PEEK slightly increased with the increase in applied loads. The influence of glass fiber on the friction coefficient and wear loss of the GF/PEEK composite is more pronounced under dry condition [7].

Bijwe and Nidhi have investigated the mechanism of adhesive wear of PEEK reinforced with GF, carbon fibers(CF) and solid lubricants (PTFE and graphite). According to these authors, the inclusion of 30 wt% carbon fibers benefited the strength properties but not the tribological performance. On the contrary, the solid lubricants influenced the friction and wear performance of PEEK composites[8]. Hamachi and Eases have studied the friction and wear of PEEK-CF30 (wt%) at elevated temperatures. The reinforcement with carbon fibers increased the mechanical resistance with the temperature increasing. As temperature increased from below to above the glass transition temperature (Tg) of the polymer matrix, the friction and wear performance of the composite slightly decreased [9].

X.X. Chu and Z.X. Wu have investigated the mechanical and thermal expansion properties of glass fibers reinforced PEEK composites at cryogenic temperatures .It was found that a dependence of mechanical properties of glass fibers reinforced PEEK composites on temperature and the thermal expansion coefficient of PEEK matrix was nearly a constant in this temperature region, and it can be significantly decreased by adding glass fibers [10].

In this work, the tribological behavior and the mechanical properties of pure PEEK and PEEK containing 30 wt% short-cutting glass fibers were studied at room temperature. SEM and EDS are used to study the strengthening and toughening mechanisms of GF/PEEK after testing, thermal decomposition of pure and GF/PEEK were also investigated by thermo gravimetric analysis.

EXPERIMENTAL

Materials

The pure PEEK and 30wt% short-cutting glass fibers reinforced PEEK composites (GF/PEEK) were supplied by Sino®-rich as injection molded $80 \times 80 mm^2$ plaques with thickness of 2mm. Test samples of dimensions approximately $20 \times 15 \times 2$ mm^3 were cut by a diamond cutter.

Mechanical Testing

Tensile and flexural test were performed according to the national standards as GB/T 1447-2005 and GB/T 1449- 2005 respectively, using JIN DAO AG-1 testing machine at room temperature. The tensile modulus and flexural modulus, tensile strength and flexural strength of the materials were measured at room temperature. Five specimens were tested for each composition to improve the reproducibility. The cross-head speeds are 10 mm/min and 5 mm/ min for tensile and flexural testing. The geometries and dimensions of the specimens for the experiments are shown in Fig. 1.

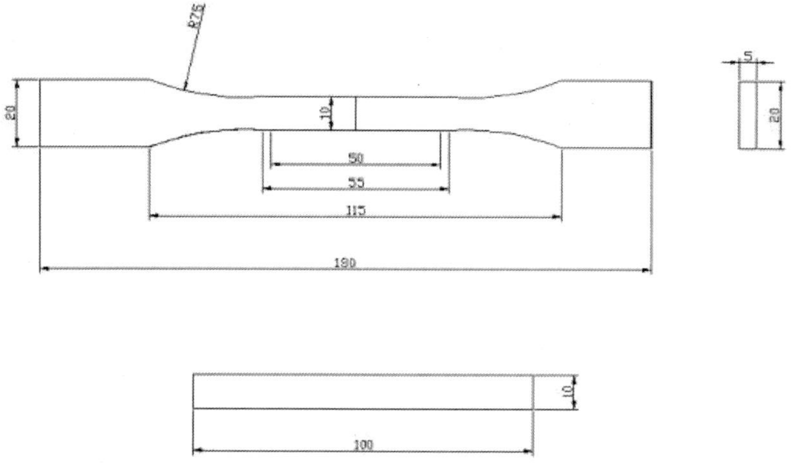

Figure 1. The geometries and dimensions of the specimens for (a) tensile and (b) flexural. Friction and wear tests

Figure. 2(a) shows the ball-on-disc wear test apparatus that was designed and used to study the tribological properties of PEEK composites. During testing, the polymeric samples were supported by a steel support (Fig. 2b). The methodology consist of a stainless steel table which is mounted on a turntable, a variable speed motor which provides the reciprocating motion to the turntable, and the steel ball which is mounted on the spring suspension is pressed against sample with the desirable load which is controlled by the press sensor. Before each test, the polymeric samples and the steel ball were cleaned with acetone. All the tests were performed under dry conditions.

The parameters used are shown in Table 1. The friction coefficient was obtained directly through the computer. The wear loss was calculated from the mass difference before and after the tribo-testing by a precision analytical balance. Each test was repeated three times. The close results were considered and their average values were presented.

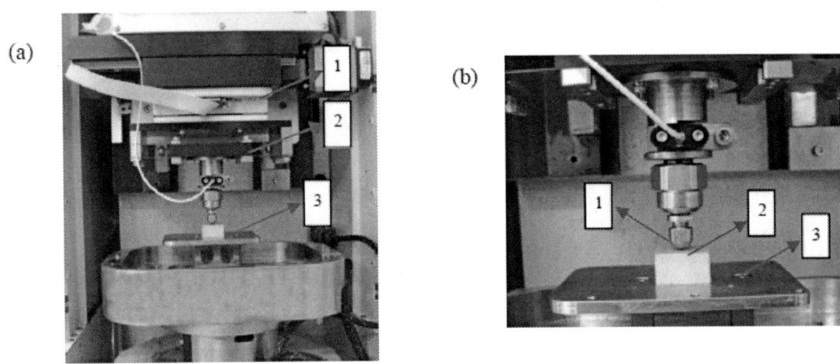

Figure 2. a Experimental set up: ball-on-disc CETR tribo-tester. 1: force sensor; 2: spring suspension; 3: ball-on-disc (b)1:ball;2:sample of polymeric material;3 stainless steel disc.

Surface Morphology Characterization

The surface morphologies of pure PEEK and GF/PEEK composites were observed by a Nova Nano SEM 450/650 scanning electron micro scope (SEM). Before SEM examination, the fracture surfaces were cleaned with alcohol. A sputtering device from Balzers SCD 050 sputter coater was used to make a conductive coating on the sample surface. An accelerating voltage of 15 kV was used for the investigation. The surface composition of the worn surfaces was determined using EDX which was coupled with SEM.

Thermogravimetric Analysis

Thermal decomposition was investigated by thermo gravimetric analysis (TGA) in an inert atmosphere using a Mettler Toledo TGA/SDTA 851. Measurements were performed on samples of 10 mg under a nitrogen atmosphere with a heating rate of 10□/min and flow rate of 50 mL/min.

Table 1. Materials and test conditions

Materials	Parameter				
	Loading stress/N	Time/min	Rate of recurrence/ Hz	Test temperature/℃	Humidity/%
Pure PEEK	100	30	2	18	56
	200	120			
	300	30			
	400	30			
GF/PEEK composites	100	30			
	200	120			
	300	30			
	400	30			

RESULTS AND DISCUSSION

Mechanical Properties

Table 2 presents the mechanical properties of pure PEEK and GF/PEEK. The tensile strength and tensile modulus values of GF/PEEK were increased by 64% and 75%, respectively. The flexural strength and flexural modules of GF/PEEK were increased by 66% and 140% compared with pure PEEK. It is clear that the reinforcing material of glass fiber strongly affected the mechanical properties of PEEK.

Table 2. The mechanical properties of pure PEEK and GF/PEEK

Sample	Pure PEEK	GF/PEEK
Tensile strength (MPa)	90.3	148.2
Tensile modulus (GPa)	2.8	4.9
Flexural strength (MPa)	139.1	230.7
Flexural modulus (GPa)	3.7	8.9

Wear Loss and Friction Coefficient

Fig. 3 presents the variation of friction coefficient for pure PEEK and GF/PEEK with the sliding time under the applied load of 200 N. An increase in the friction coefficient is observed with the increase in sliding time. Initially, a sharp increase of friction coefficients indicating the initial running-in period was observed. Then the friction coefficient tended to be stable. The friction coefficient of GF/PEEK was higher than that of pure PEEK throughout the whole test.

Fig.4 (a) shows the variation of friction coefficient for pure PEEK and GF/PEEK with the change of applied load. The friction coefficient of PEEK increased steadily with the increase of applied load. The friction coefficient increased by about 60% when the load changed from 200 N to 300 N. The friction coefficient of GF/PEEK decreased initially and then slightly increased with the applied load. It is well known that PEEK polymer is a viscoelastic material, so the variation of friction coefficient with the load follows the equation $\mu = K N^{(n-1)}$, where μ is the friction coefficient, N is the applied load, K is a constant and n is also a constant with values between 2/3 and 1[11].

According to this equation, the friction coefficient decreases with the load increasing. However, when the load increases to the critical load of the pure PEEK, the friction and wear will increase sharply. This behavior is because that the friction heat raised the temperature of friction surfaces, which leads to relaxation of polymer molecule chains.

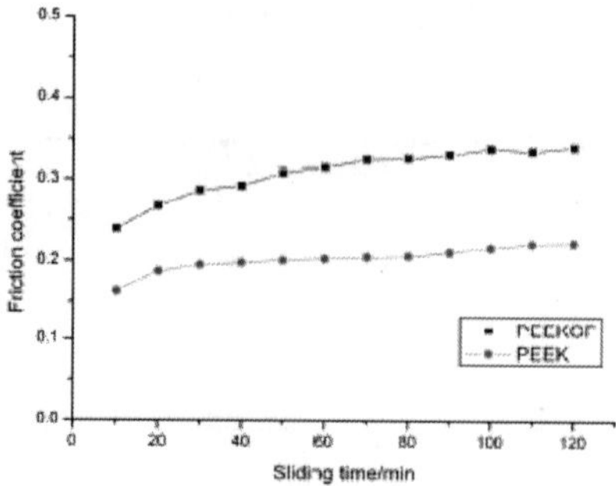

Figure 3..Variation of the friction coefficient for the PEEK and GF/PEEK, applied load=200 N

Figure. 4(b) shows wear losses of PEEK and GF/PEEK under different applied load. A linear increase of the weight loss is observed as the increasing of the applied load. The increase for the PEEK is more noticeable than that for GF/PEEK, indicating GF/PEEK has greater wear resistance than PEEK.

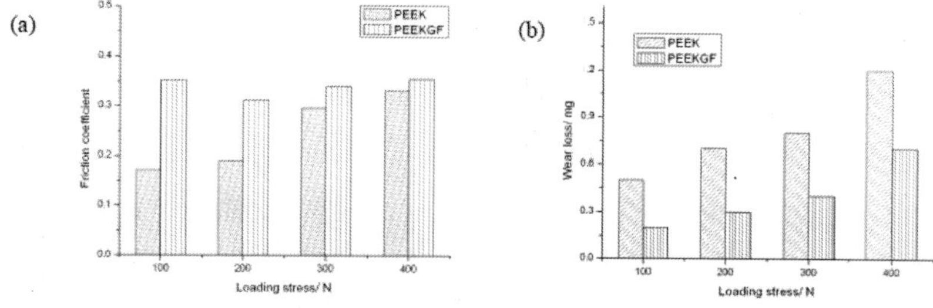

Figure 4. The variation of friction coefficient and wear loss with applied load of pure PEEK and GF/PEEK, sliding time=30min.

Worn Surface Analysis

The worn surfaces were observed by scanning electron microscopy(SEM), to identify the wear mechanisms. Fig.5 shows the worn surface morphologies of PEEK. It can be seen from the Fig.5(a) that the wear debris of PEEK adhered to the matrix of polymer. The formation and adhesion of the wear debris are typical features of the wear process of PEEK[12]. Fig.5(b) shows a higher magnifying view(5000×).It can be observed that the PEEK wrapped into together. This is explained as under grinding time longer generates heat at the polymer and steel counterface contact area, the PEEK polymer surface is softening and as a result of increasing in friction coefficient and wearloss.

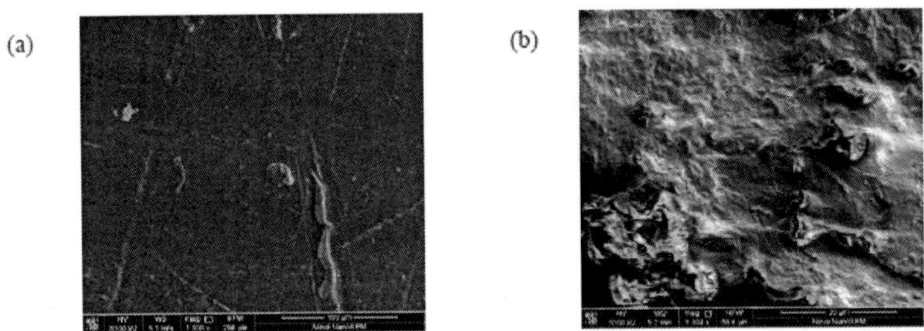

Figure 5. SEM pictures of pure PEEK under the applied load of 200N and sliding time of 30min

Figure. 6 show the worn morphologies of GF/PEEK composites. The glass fibers were dispersed disorderly in the matrix. Short pull-out lengths of the fibers were the evidence of strong fiber/matrix interfacial bond. Fig. 6(b) illustrates the worn surface of the GF/PEEK where the material removal of the matrix is interrupted by the fibers, causing the accumulation of wear debris around the glass fiber (indicated by arrow) [13].

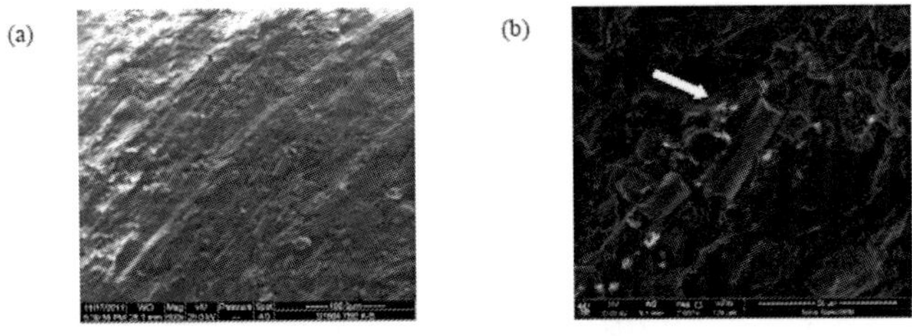

Figure 6. SEM pictures of GF/PEEK under 200N applied load and sliding time of 30min

Figure. 7(a) illustrates the low magnification worn surface of the GF/PEEK under the applied load of 400N and the sliding time of 120min. An obvious boundary was produced between the worn area and the unworn area (indicated by the red line). The worn area was very rough which may be due to the compatibility between PEEK and short glass fibers, or caused by the molding process. A smoother surface is observed in the worn area compared with the unworn area. Fig. 7(b) is medium magnification SEM of the boundary between the worn area and unworn area. Part of the matrix material was squeezed out of the surface due to the compressive load and no glass fibers included (similar to the process which was observed under single point scratch studies with the same material[14]) covered the unworn areas, providing temporarily an additional protection of the underlying fiber ends, it was associated with a high degree of fiber/ matrix debonding. The microanalysis for X-ray(EDS) showed that the short glass was extruded from the composites by the applied load rather than pulverized into the composite. The EDS of worn and unworn area are given in Fig. 7(c) and 7(d). The debris that covered the unworn area only contained carbon and oxygen element(Spectrum 1). The short glass fibers contained silicon, calcium and aluminum element(Spectrum 2). Fig. 7(e) shows a higher magnifying view (5000×), the remained glass fibers well embedded in the polymer matrix and the matrix material around the fibers was disrupted. The short glass fibers significantly improved the triblogical and mechanical properties of GF/PEEK composites.

Thermogravimetric Analysis

In general, introduction of the filler increases the thermal stability owing to high temperature degradation as

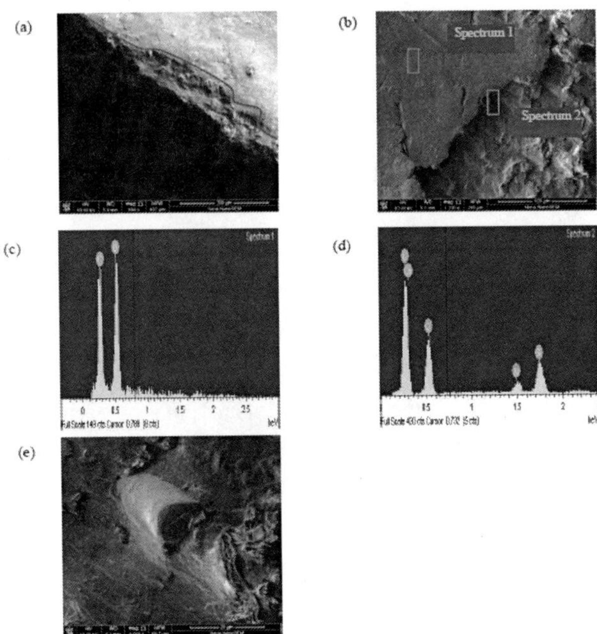

Figure 7. Surface damage mechanisms of a short glass fiber reinforced PEEK at 200N applied load. a low magnification of wear, b medium magnification, c the EDS of spectrum1, d the EDS of spectrum2, e higher magnification(5000×).

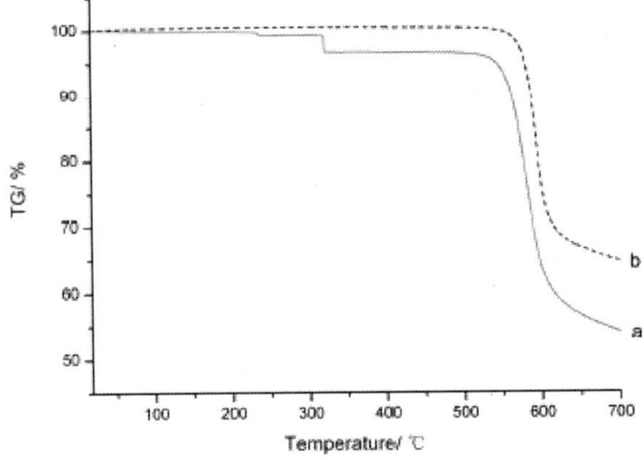

Figure 8. the typical TG weight loss curves for the pure PEEK and GF/PEEK

CONCLUSION

The mechanical properties, tribological behavior and thermal properties of pure PEEK and 30wt% short glass fibers reinforced PEEK prepared have been studied in this paper, considering the used methodology:

- The tensile strength, tensile modulus, flexural strength and flexural modulus of GF/PEEK were higher than those of pure PEEK, the incorporation of short glass fibers into PEEK polymer matrix obviously improved the mechanical performance.
- For all the tested materials, the friction coefficient and wear loss increased with the increase of the sliding time and the applied load. The GF/PEEK presented better wear resistance than PEEK.
- The bond between short glass fiber and the PEEK was strong fiber/matrix interfacial bond. The short glass fibers were extruded from the composite rather than pulverized into the composite.
- The weight loss temperatures of GF reinforced composites were about 75□ higher than that of pure PEEK, the incorporation of short glass fiber into the PEEK polymer matrix increased the thermal stability.

ACKNOWLEDGEMENTS

The paper is financially supported by Innovation Fund of Academy of Armored Forces Engineering(2013CJ36) ,National Key Basic Program of China 973 (2011CB013403) and Distinguished Youth Scholars of National Natural Science Foundation of China (51125023).

REFERENCES

1. Parina Patel, T. Richard Hull, Richard E. Lyon, et al. Investigation of the thermal decomposition and flammability of PEEK and its carbon and glass-fiber composites[J]. Polymer Degradation and Stability, 2011(96):12-22.
2. G.Y. Xie, G.X. Sui, R. Yang. Effects of potassium titanate whiskers and carbon fibers on the wear behavior of polyetheretherketone composite under water lubricated condition[J]. Composites Science and Technology, 2011(71):828-

835.

3. Avci A, Arikan H, Akdemir A. Fracture behavior of glass fiber reinforced polymer composite[J]. Cem Concr Res, 2004(34):429-34.
4. T.C.Ovaert, H.S.Cheng. Counterface topographical effects on the wear of polyetheretherketone and a polyetheretherketone-carbon fiber composite[J]. Wear, 1991(150):150-157.
5. Shang LG. Characterisation of interphase nanoscale property variations in glass fibre reinforced polypropylene and epoxy resin composites[J]. Composites Part A 2002, 33(4):559-76.
6. Z.P.Lu, K.Friedrich. On sliding friction and wear of PEEK and its composites[J]. Wear, 1995:624-631.
7. M.Sumer, H.Unal, A.Mimaroglu. Evaluation of tribological behavior of PEEK and glass fibre reinforced PEEK composite under dry sliding and wear lubricated conditions[J]. Wear, 2008(265):1061 1065.
8. J. Bijwe, Nidhi. Potential of fibres and solid lubricants to enhance the triboutility of PEEK in adverse operating conditions[J]. Industrial Lubrication Tribology, 2007 (4) :156-165.
9. J. Hanchi, N.S. Eiss Jr. Dry sliding friction and wear of short carbonfiber-reinforced polyetheretherketone (PEEK) at elevated temperatures[J]. Wear, 1997 (203-204): 380-386.
10. X.X. Chu, Z.X. Wu, R.J. Huang, et al. Mechanical and thermal expansion properties of glass fibers reinforced PEEK composites at cryogenic temperatures[J]. Cryogenics, 2010(50):84-88.
11. H.Uetz, J.Wiedemeyer. Trbologie der Polymere, Carh hanser Verlag, Munchen Vienna, 1985.
12. H.Voss, K.Friedrich. On the wear behavior of short-fiber-reinforced PEEK composites. [J].wear, 1987,116:1-18.
13. J.Paulo Davim, Rosaria Cardoso. Effect of the reinforcement(carbon or glass fibers) on friction and wear behavior of the PEEK against steel surface at long dry sliding [J].wear,2009,266:795-799.
14. Friedrich, K., Goda, T., Varadi, K., Wetzel, B.: FE-Simulation of the fiber/ matrix debonding in polymer composites produced by a sliding indentor, Part 1: Normally oriented fibers[J]. J. Compos. Mater. 2004,38, 1583- 1606.

Chapter 10

SYNTHESIS AND CHARACTERIZATION OF MECHANICAL PROPERTIES IN COTTON FIBER-REINFORCED GEOPOLYMER COMPOSITES

T. Alomayri, I.M. Low

Department of Imaging & Applied Physics, Curtin University, GPO Box U1987, Perth, WA 6845, Australia

ABSTRACT

Geopolymers are inorganic aluminosilicate materials that possess relatively good mechanical properties and desirable thermal stability but they exhibit failure behavior similar to brittle solids. This limitation may be remedied by fiber reinforcement to improve their strength and toughness. This paper describes the synthesis of cotton fiber-reinforced geopolymer composites and the characterization of their mechanical properties. The effects of cotton fiber content (0–1.0 wt.%) and fiber dispersion on the mechanical characteristics of geopolymer composites have been investigated in terms of hardness, impact strength and compressive strength. A fiber content of 0.5 wt.% was observed for achieving optimum mechanical properties in these composites.

INTRODUCTION

Inorganic aluminosilicate Portland cements are used in many building and construction applications because of their good mechanical performance. However, the emission of greenhouse gases associated with their manufacture is a serious problem. In recent years, a new class of environment-friendly and sustainable inorganic aluminosilicate polymers (also known as geopolymers) has emerged as an alternative to cements. These inorganic compounds can be cured and hardened at near-ambient temperatures to form materials that are effectively low-temperature ceramics with the typical temperature resistance and strength of ceramics [1] and [2]. However, despite their many desirable attributes such as relatively high strength, elastic modulus and low shrinkage, geopolymers suffer from brittle failure like most ceramics. This limitation may be readily overcome through fiber reinforcement as in high performance polymer-matrix composites. As in thermosetting polymers, the low synthesis temperatures of geopolymers renders them particularly suitable as matrices for a range of fibers including organic fibers, with setting times and mechanical properties comparable to Portland cement [3]. Hitherto, the most common fiber reinforcement used in geopolymer composites is based on carbon, basalt and glass fibers, but other inorganic fibers such as silicon carbide, alumina, mullite or boron can be utilized [4], [5] and [6]. Maximum flexural strengths of >500 MPa have been reported by several authors for unidirectional carbon fiber-reinforced geopolymer composites [4] and [7] and desirable non-brittle fracture was observed when short carbon fibers were used [8].

Current concerns over the environment and climate change have also given rise to an increasing interest in replacing the synthetic fibers currently used in geopolymer composites or other brittle matrices with natural plant fibers. Plant fibers cost less, have low density and display good mechanical properties when compared with industrial fibers [9], [10] and [11]. Investigations on natural fibers such as bamboo, sisal, jute and cellulose have revealed desirable effects on the mechanical and physical properties of brittle organic and inorganic matrices. For instance, the mechanical and fracture properties of epoxy resin have been significantly improved as a result of cellulose fiber reinforcement [12], [13] and [14]. Similarly, Rahman et al. [15] found that bamboo fibers are effective in improving the flexural strength of concrete, and Lin et al. [16] observed a similar desirable effect in wood fiber-reinforced concrete. In another study, Li et al. [17] found that hemp fibers enhanced the toughness of concrete. Wool fibers have also been successfully used in reinforcing geopolymer composites with concomitant improvements in mechanical and fracture properties [18]. However, the use of cotton fibers as reinforcement for geopolymers has not been investigated. The use of cotton fibers has several advantages, which include low cost, renewable, and low weight when compared with synthetic fibers.

In this paper, we have synthesized geopolymer composites reinforced with short cotton fibers and characterized their mechanical properties in terms of hardness, compressive strength and impact strength. The effect of fiber contents (0.3, 0.5, 0.7 and 1 wt.%) and their dispersion on mechanical properties were investigated. Scanning electron microscopy (SEM) was used to examine the microstructures of fly-ash and the resultant composites.

EXPERIMENTAL PROCEDURE

Materials

Low calcium fly-ash (ASTM class F) [19] collected from the Collie power station in Western Australia was used as the source material to prepare the geopolymer composites. The chemical compositions of fly-ash are given in Table 1. Alkali resistant cotton fibers with an average length of 10 mm, average diameter of 0.2 mm, density of 1.54 g/cm3, tensile strength of 400 MPa, and Young's modulus of 4.8 GPa were used to reinforce the geopolymer composites. The alkaline activator for geopolymerization was a combination of sodium hydroxide solution and sodium silicate grade D solution. Sodium hydroxide flakes with 98% purity were used to prepare the solution. The chemical composition of sodium silicate is Na2O 14.7%, SiO2

29.4% and water 55.9% by mass.

Table 1. Chemical composition of fly-ash.

SiO_2	Al_2O_3	Fe_2O_3	CaO	MgO	SO3	Na2O	K2O	LOI
50%	28.25%	13.5%	1.78%	0.89%	0.38%	0.32%	0.46%	1.64%

SAMPLE PREPARATION

To prepare the geopolymer composites, an alkaline solution to fly-ash ratio of 0.35 was used and the ratio of sodium silicate solution to sodium hydroxide solution was fixed at 2.5. Four samples of geopolymer composites reinforced with 0.3, 0.5, 0.7 and 1 wt.% cotton fibers were prepared. Additional water was added to improve the workability and dispersion of cotton fibers in the composites.

An 8 M concentration of sodium hydroxide solution was prepared and it was combined with the sodium silicate solution 1 day before mixing. The fibers were added slowly to the dry fly-ash in a Hobart mixer at a low speed until the mix became homogeneous at which time the alkaline solution was added. This was mixed for 10 min on low speed and for another 10 min on high speed. The walls of the mixing container were scraped down to ensure consistency of the mix. This procedure was followed for all the four test specimens. The mix was cast in 25 rectangular silicon molds of 80 mm × 20 mm × 10 mm and placed on a vibration table for 5 min. The specimens were covered with a plastic film and cured at 105 °C for 3 h, then rested for 24 h before demolding. They were then dried under ambient conditions for 28 days.

Synchrotron Radiation Diffraction (SRD)

The Powder Diffraction beamline at the Australian Synchrotron was used to collect the diffraction patterns of fly-ash, cotton fibers and the geopolymer composites. The diffraction pattern of each sample was collected using an incident angle of 3° and wavelength of 0.11267 nm or photon energy of 11.0 keV over the 2θ range of 10°–40°

Scanning Electron Microscopy (SEM)

A Zeiss Evo 40XVP scanning electron microscope was used to examine the microstructures of fly-ash and geopolymer composites. The specimens were mounted on aluminum stubs using carbon tape, and then coated with a thin layer of platinum to prevent charging before the observation.

Rockwell Hardness

The hardness of geopolymer composites was measured using an Avery Rockwell hardness tester at hardness scale H. Before measurement, the surfaces of test samples were polished using a Struers Pedamat polisher finishing with 10-μm grade diamond paste.

Compressive Strength

The measurement of compressive strength testing was conducted using the methodology of ASTM C39 for concrete specimens. Cylindrical samples with a 2:1 height to diameter ratio were cut with a precision diamond blade such that the ends were perpendicular to the sides. A minimum of five samples were tested. Following demolding, the samples were air dried for 1 day before the compressive test. An EZ50 (Lloyd Instruments Ltd., West Sussex, UK) was used to apply a constant stress rate of 0.25 MPa/s, after a 50 N preload, until failure

$$C = \frac{P}{A}$$

where P is total load on the sample at failure and A is calculated area of the bearing surface of the specimen

Impact Strength

Rectangular bars with dimensions 80 mm × 20 mm × 10 mm were prepared for Zwick Charpy impact testing to evaluate the impact strength of geopolymer composites. A pendulum hammer with 1.0 J was used during the test to break the samples. Un notched samples were used to compute the impact strength (i) using the following formula:

$$\sigma_i = \frac{E}{A}$$

where E is the impact energy to break a sample with a ligament of area A

RESULTS AND DISCUSSION

Synchrotron Radiation Diffraction

The synchrotron radiation diffraction (SRD) patterns of commercial fly-ash, cotton fibers and prepared geopolymer reinforced with 0.3, 0.5, 0.7 and 1.0 wt.% of cotton fibers are shown in Fi , indicating the presence of cellulose. Fly-ash displays peaks due to the presence of quartz and mullite as well as other crystalline phases. These crystalline phases are not involved in the geopolymerization reaction, but the amorphous phase generated by coal combustion is actively involved in geopolymerization reac tions [20]. Rickard et al. [21] have recently shown that amorphous aluminosilicates in fly-ash are reactive during the formation of a geopolymer.

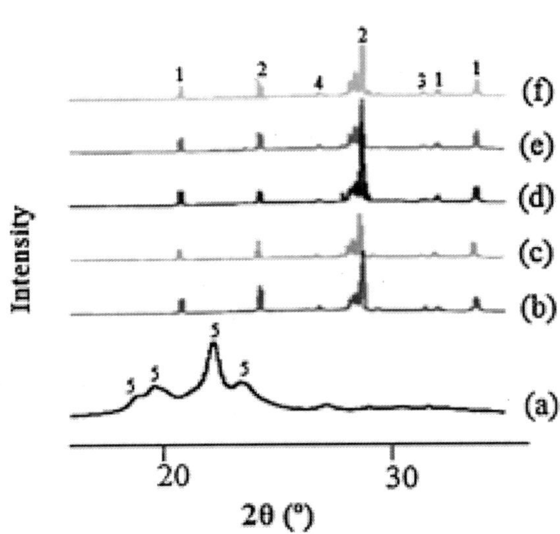

Figure. 1. Synchrotron radiation diffraction patterns of (a) cotton fibers (CF), (b) fly ash, and geopolymer composite with (c) 0.3 wt.% CF, (d) 0.5 wt.% CF, (e) 0.7 wt.% CF, and (f) 1.0 wt.% CF. [Legend: 1 = mullite, 2 = quartz, 3 = maghemite, 4 = hematite, 5 = cellulose].

Comparing the SRD spectra of the original fly-ash with those of the hardened geopolymeric materials (see Fig. 2) indicates that the crystalline phases (quartz, mullite, etc.) originally existing in the fly-ash have apparently not been altered by the activation reactions; hence they do not participate in the geopolymerization reaction. The diffraction patterns of geopolymer reinforced with 0, 0.3, 0.5, 0.7 and 1 wt.% cotton fibers showed the sharp peaks of the crystalline phases from fly-ash, thus confirming that these phases are neither reactive nor involved in geopolymerization but are simply present as inactive fillers in the geopolymer network

SEM Observation

The SEM micrographs of fly-ash and geopolymer composites loaded with fiber content of 0.5 wt.% are shown in Figs. 2 and 3. Figure. 2 shows the microstructure of the original fly-ash before being activated with the alkaline activator. As seen in the figure, the fly ash consists of spherical particles of different sizes. Some particles may contain smaller particles in their interior [22]. The surface tex

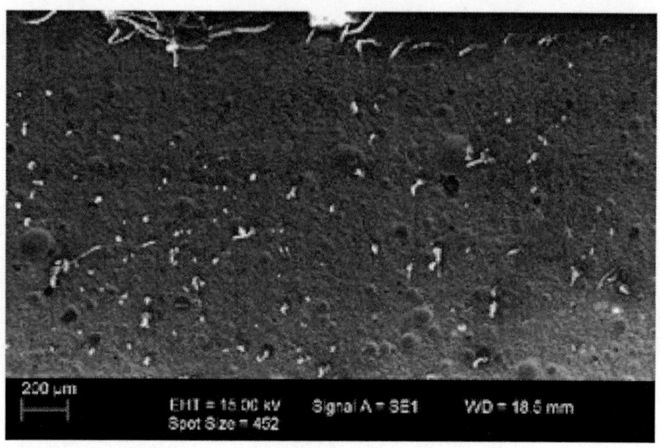

Figure. 3. SEM micrograph showing the typical microstructure of geopolymer composite reinforced with 0.5 wt.% cotton fibers. SEM, Scanning electron microscopy.

Fig. 3 shows that at 0.5 wt.% cotton fiber, the fibers are distributed homogeneously within the matrix. The uniformity of cotton fiber distribution in the matrix plays crucial roles in governing the properties of the composites. To gain advantageous properties, the following factor should be considered during fabrication of cotton fiber-reinforced geopolymer composites.

Hardness of Geopolymer Composites

The effect of cotton fiber content on the hardness of the cotton fiber-reinforced geopolymer composites is presented in Fig. 4. The hardness of geopolymer reinforced with 0.5 wt.% cotton fiber increased from 70 to 93 Rockwell hardness H (HRH) relative to the neat geopolymer. This significant enhancement in hardness is due to distribution of the test load on the fibers, which decreased the penetration of the test ball to the surface of the composite material and consequently raising the hardness of this material [25].

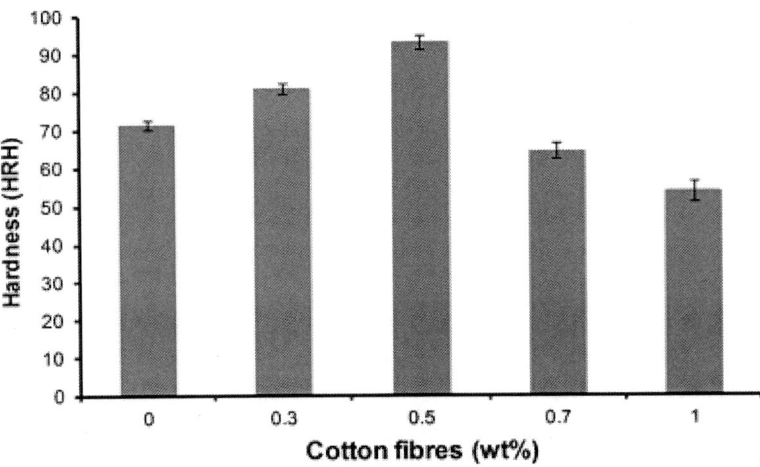

Figure. 4. Hardness of geopolymer composites as a function of fiber content.

However, the hardness decreased with increasing fiber content due to the poor dispersion of cotton fibers in the slurry. The addition of 0.7 and 1.0 wt.% cotton fibers resulted in a reduction in the consistency of the matrix as well as low wettability between the fibers and the paste, and the fibers could be separated from the paste easily. This had to be compensated for by an increase in the water content of the mix. Increasing water content to overcome such a problem may lead to low hardness. The research conducted by Kunal [26]revealed that higher water content results in samples with low hardness. Because a higher than normal water content was needed for the samples to be flexible, the strength of the samples was reduced.

Similarly, this decrease has been reported by other researchers when dealing with natural fiber based composites. Anup [27] reported that with increasing flax fiber content, the hardness value of high-density polyethylene/flax fiber composites and polypropylene/flax fiber composites decreased. Khairaih and Khairul[28] also reported decreasing hardness values with increasing fiber content when they worked on polyurethane and empty fruit bunch blend composites. They concluded that the decrease was due to the inability of the matrix to encapsulate the fiber strands.

Compressive Strength of Geopolymer Composites

The 28-day average values of compressive strength of the composites are given in Fig. 5 and their corresponding stress/strain curves are shown in Fig. 6. It can be seen that geopolymer composite with 0.5% cotton fibers had the highest compressive strength. This is attributed to the possibility that the higher loads transferred from the matrix to the fibers, thus resulting in a higher load carried by the fibers. Another reason for such favorable behavior could be good dispersion of cotton fibers throughout the matrix that increases the bonding strength between the fiber and the matrix. From the stress–strain curves in Fig. 6, it is interesting to note that geopolymer composites displayed some non-linearity during fracture whereas a linear fracture behavior was observed for geopolymer. This implies the feasibility of using cotton fibers to mitigate the brittle failure in geopolymers.

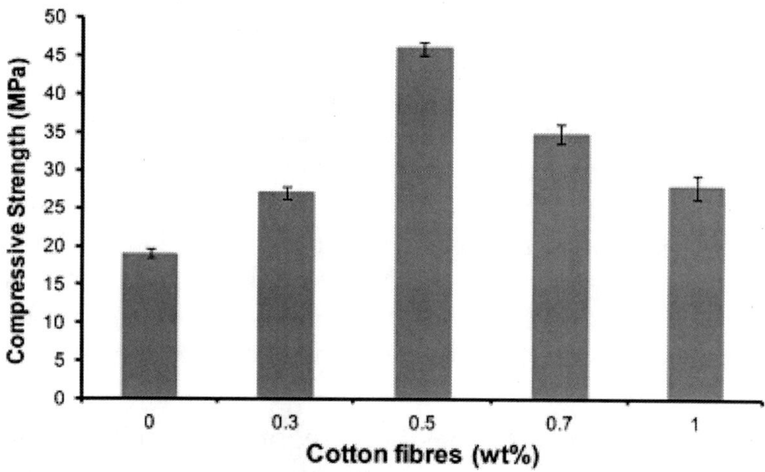

Figure. 5. Compressive strength of geopolymer composites as a function of fiber content.

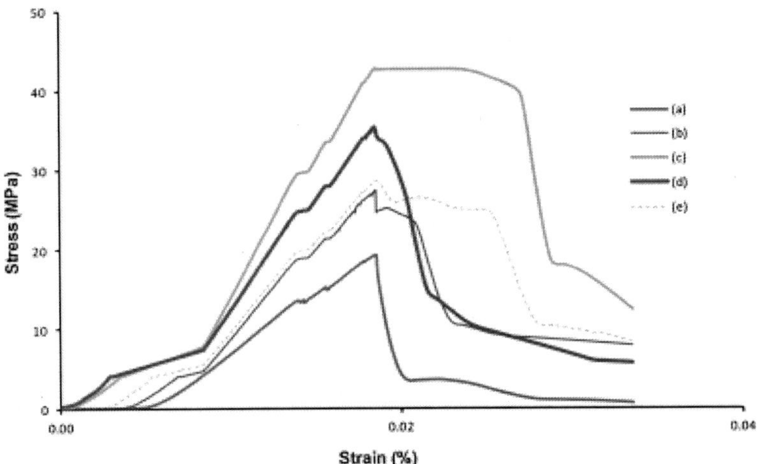

Figure. 6. Typical stress–strain curves of geopolymer composites with various cotton fiber contents (a) 0 wt.%, (b) 0.3 wt.%, (c) 0.5 wt.%, (d) 0.7 wt.%, and (e) 1.0 wt.%.

However, the geopolymer composites cast with cotton fiber in the amount of about 0.7 and 1% fiber content by weight yielded a weak compressive strength. The reason for the reduction in compressive strength instead of an improvement with the addition of cotton fibers may be attributed to a greater probability of these fibers balling together and leaving voids in the matrix [29].

Other reasons for this weakness may be that the cotton fibers had absorbed too much water, denying the geopolymer around the fibers enough water for geopolymerization, which in turn decreased the bonding strength between the fiber and the matrix.

Similar results were reported by Li et al. [17], who investigated the compressive properties of hemp fiber-reinforced concrete. They found that compressive strength improves slightly when the fiber content by weight is lower than 0.6%, and continuously decreases when the fiber content is greater than this value.

In the present study, the compressive strength of the neat geopolymer paste increased from 19.1 to 46.0 MPa after the addition of 0.5 wt.% cotton fibers. However, adding more cotton fibers (0.7 and 1.0 wt %) led to a reduction in compressive strength.

Impact Strength of Geopolymer Composites

The impact strength of fiber-reinforced polymer is governed by the matrix fiber interfacial bonding, and the properties of the matrix and the fibers. When the composites undergo a sudden force, the impact energy is dissipated by the combination of fiber pull out, fiber fracture and matrix deformation [30].

The effects of fiber content on the impact strength of cotton fiber-reinforced geopolymer composites are plotted in Fig. 7. It can be seen that the impact strength of the composites increases with an increase in cotton content of up to 0.5 wt.%, and then it decreases thereafter. The enhancement in impact strength may be ascribed to the good dispersion of cotton fibers throughout the matrix, which helps to increase the interaction or adhesion at the matrix/cotton fiber interface. In addition, the increases in impact strength as fiber content increases are due to the increase in fiber pull out and fiber breakage [31]. Hence, this permits the optimum operation of stress-transfer from the matrix to the cotton fibers, thus resulting in an improvement of strength properties.

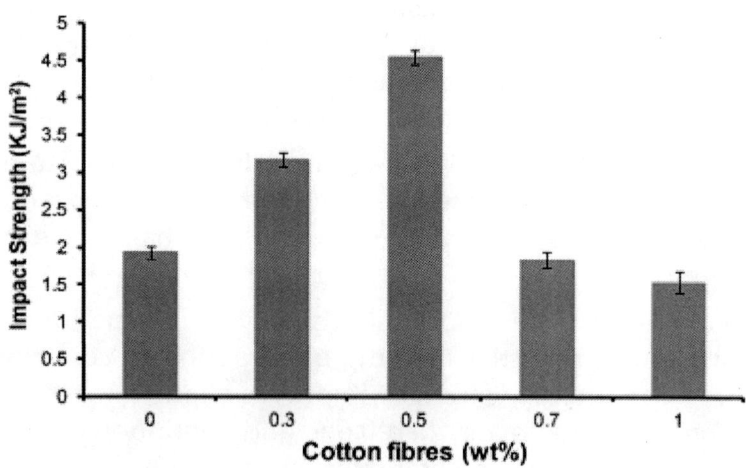

Figure. 7. Impact strength of geopolymer composites as a function of fiber content.

However, the impact strength of composites decreases when fiber content increases to >0.5 wt.%. This reduction in impact strength at higher content of cotton fiber was due to the formation of fiber agglomerates and voids as a result of increased system viscosity due to the presence of the cotton fiber, which in turn reduced the fiber matrix adhesion.

The impact strength of the neat geopolymer paste increased from 1.9 to 4.5 kJ/m^2 after the addition of 0.5 wt.% cotton fibers. However, adding more cotton fibers (0.7 and 1.0 wt %) led to a reduction in strength.

CONCLUSIONS

Geopolymer composites reinforced with cotton fibers have been fabricated and characterized. Optimum enhancements in hardness, compressive strength and impact resistance were achieved for composites containing up to 0.5 wt.% cotton fibers. However, further increase in cotton fiber content beyond 0.5 wt.% led to fiber agglomerations with a concomitant reduction in mechanical properties by virtue of increased viscosity, voids formation and poor dispersion of fibers within the matrix.

ACKNOWLEDGEMENTS

The authors would like to thank Ms E. Miller for assistance with SEM, and both Dr W. Rickard and Mr L. Vickers for assistance in mechanical tests. The collection of diffraction data was funded by the Australian Synchrotron (PD 5341).

REFERENCES

1. J. Davidovits J. Therm. Anal., 37 (1991), pp. 1633–1656
2. V.F.F. Barbosa, K.J.D. Mackenzie, C. Thurmaturgo Int. J. Inorg. Mater., 2 (2000), pp. 309–317
3. K.J.D. MacKenzie, in "Advances in Ceramic-Matrix Composites", Ed. by I.M. Low, Woodhead Publishing, Cambridge (in press, Chapter 18), ISBN: 9780857091208.
4. J.A. Hammell, P.N. Balagurn, R.E. Lyon Composite B, 31 (2000), pp. 107–111
5. E. Rill, D.R. Lowry, W.M. Kriven Mater. Comput. Des., 31 (2010), pp. 57–67
6. K. Vijai, R. Kumuthaa, B.G. Vishnuram Asian J. Civ. Eng., 13 (2011), p. 9

7. A. Foden, Ph.D. Thesis, Rutgers State University, USA (1999).
8. T. Lin, D. Jia, P. He, M. Wang, D. Liang Mater. Sci. Eng. A, 497 (2008), pp. 181–185
9. G. Ramakrishna, T. Sundararajan Cem. Concr. Compos., 27 (2005), pp. 547–553
10. D.C. Teo, M.A. Mannan, V.J. Kurian J. Adv. Concr. Technol., 4 (2006), pp. 1–10
11. R.D. Filho, K. Ghavami, G.L. England, K. Scrivener Cem. Concr. Compos., 25 (2003), pp. 185–196
12. I.M. Low, M. McGrath, D. Lawrence, P. Schmidt, J. Lane, B.A. Latella, K.S. Sim Composite A, 38 (2007), pp. 963–974
13. I.M. Low, J. Somers, H.S. Kho, I.J. Davies, B.A. Latella Compos. Interfaces, 16 (2009), pp. 659–669
14. H. Alamri, I.M. Low, Z. Alothman Composite B, 43 (2012), pp. 2762–2771
15. M.M. Rahman, M.H. Rashid, M.A. Hossain, M.T. Hasan, M.K. Hasan Int. J. Eng. Technol., 11 (2011), pp. 142–146
16. X. Lin, M.R. Silsbee, D.M. Roy, K. Kessler, P.R. Blankenhorn Cem. Concr. Res., 24 (1994), pp. 1558–1566
17. Z. Li, L. Wang, X. Wang Fibers Polym., 5 (2004), pp. 187–197
18. M.K. Alzeer, K.J.D. MacKenzie J. Mater. Sci., 47 (2012), pp. 6958–6965
19. ASTM C39 (2005).
20. U. Rattanasak, P. Chindaprasirt Miner. Eng., 22 (2009), pp. 1073–1078
21. W.D.A. Rickard, R. Williams, J. Temuujin, A. Van Riessen Mater. Sci. Eng. A., 528 (2011), pp. 3390–3397
22. A. Fernandez-Jimenez, A. Palomo, M. Criado Cem. Concr. Res., 35 (2005), pp. 1204–1209
23. A. Fernandez-Jimenez, A. Palomo Fuel, 82 (2003), pp. 2259–2265
24. A. Fernandez-Jimenez, A. Palomo Cem. Concr. Res., 35 (2005), pp. 1984–1992
25. H.P. Abbasi, M.Sc. Thesis, King Fahd University of Petroleum and Minerals, Saudi Arabia (2003).
26. S. Kunal, M.Sc. Thesis, Rochester Institute of Technology, New York (2006).
27. R. Anup, M.Sc. Thesis, University of Saskatchewan (2008).
28. A. Khairiah, A. Khairul J. Oil Palm Res. (2006), pp. 103–113
29. D.P. Dias, C. Thaumaturgo Cem. Concr. Compos., 27 (2005), pp. 49–54
30. P. Wambua, J. Ivens, I. Verpoest Compos. Sci. Technol., 63 (2003), pp. 1259–1264
31. S. Mishra, A.K. Mohanty, L.T. Drzal, M. Misra, S. Parija Compos. Sci. Technol., 63 (2003), pp. 1377–1385

Chapter 11

SELF-HEALING CAPABILITY OF FIBER-REINFORCED CEMENTITIOUS COMPOSITES FOR RECOVERY OF WATERTIGHTNESS AND MECHANICAL PROPERTIES

Tomoya Nishiwaki [1,*], Sukmin Kwon [1], Daisuke Homma [2], Makoto Yamada [1] and Hirozo Mihashi [1]

[1] Department of Architecture and Building Science, Tohoku University, Aoba 6-6-11-1205, Aramaki, Aoba-ku, Sendai 980-8579, Japan; E-Mails: sukminkwon@hjogi.pln.archi.tohoku.ac.jp (S.K.); makotoyamada@hjogi.pln.archi.tohoku.ac.jp (M.Y.); h-mihashi@ab.cyberhome.ne.jp (H.M.)

[2] Takenaka Research & Development Institute, Takenaka Corporation, Ohtsuka 1-5-1, Inzai, Chiba 270-1395, Japan; E-Mail: honma.daisuke@takenaka.co.jp

* Author to whom correspondence should be addressed; E-Mail: ty@archi.tohoku.ac.jp; Tel.: +81-22-795-7864; Fax: +81-22-795-4772.

ABSTRACT

Various types of fiber reinforced cementitious composites (FRCCs) were experimentally studied to evaluate their self-healing capabilities regarding their watertightness and mechanical properties. Cracks were induced in the FRCC specimens during a tensile loading test, and the specimens were then immersed in static water for self-healing. By water permeability and reloading tests, it was determined that the FRCCs containing synthetic fiber and cracks of width within a certain range (<0.1 mm) exhibited good self-healing capabilities regarding their watertightness. Particularly, the high polarity of the synthetic fiber (polyvinyl alcohol (PVA)) series and hybrid fiber reinforcing (polyethylene (PE) and steel code (SC))series showed high recovery ratio. Moreover, these series also showed high potential of self-healing of mechanical properties. It was confirmed that recovery of mechanical property could be obtained only in case when crack width was sufficiently narrow, both the visible surface cracks and the very fine cracks around the bridging of the SC fibers. Recovery of the bond strength by filling of the very fine cracks around the bridging fibers enhanced the recovery of the mechanical property.

INTRODUCTION

Significant effort toward the development of a sustainable society has been made in recent times. The maintenance of concrete structures has become one of the most important issues from the ecological point of view. In Japan, several concrete infrastructures built about 50 years ago have deteriorated. Comprehensive maintenance is therefore necessary. However, there has been a decrease in the population of Japan, which has led to the pressing need for labor-saving and cost-effective maintenance systems. Moreover, the functional and/or environmental conditions of some infrastructures such as urban highways and facilities for disposal of radioactive waste make accessibility by engineers for inspection and maintenance difficult or impossible.

The adoption of self-healing methods is one of the most promising approaches to dealing with inherent cracks in concrete structures without human intervention. Many studies have been conducted on the self-healing capability of concrete, especially in the last decade [1–3]. However, a practical solution is still not available. Several studies are ongoing, mainly in Europe, regarding a bacterial approach, e.g., [4,5]. This novel method involves the biological precipitation of calcium carbonate in the cracks through the reaction of CO_2 produced by bacterial metabolism with $Ca(OH)_2$ present in the cement matrix. However, the survival of bacteria has been a challenge, because they are killed during the mixing and setting of concrete, not to talk of long-term survival over the lifetime of concrete structures. Other methods, including the embedment of healing agents (adhesives) contained by capsule and/or vascular approaches, have been investigated [6–11]. In these cases, the pre-maturation of the adhesives before crack initiation and the procedures (mixing and setting of the concrete using brittle pipes and/or capsules) are the challenges that need to be overcome. Moreover, the additional components or devices may be detrimental to the concrete structure. In the vascular method, it is very difficult to construct a network that covers the entire structure like the blood vessels of a human body.

Fiber reinforced cementitious composites (FRCCs) have the greatest potential for the practical achievement of self-healing in concrete. FRCCs are currently commonly used with good results in real applications. Because no special components are required, durability and cost efficiency are assured. Ubiquitous reinforcing fibers in the cement matrix can be used to enhance the self-healing by controlling the crack width and acceleration of $CaCO_3$ precipitation [12,13]. Because the healing process does not involve the reaction of the fiber itself, the problem of ensuring the durability of the healing agent is solved. Reinforcement fibers containing high-polarity synthetic composite (e.g., PVA) particularly have strong potentials for self-healing precipitation around the fibers that bridge a crack [14].

In this study, various types of FRCCs were experimentally studied to evaluate their self-healing capabilities regarding their water permeability and mechanical properties.

Experimental Program

Materials and Mixture Proportions

The mixture proportions of the FRCCs are summarized in Table 1. High-early-strength Portland cement and silica fume were used as the binder. The densities of the cement and silica fume were 3.14 g/cm^3 and 2.2 g/cm^3 respectively. The aggregates comprised well-graded and very fine natural silica sand with an average particle size of 180 μm. A polycarboxylate-based superplasticizer was employed. The properties of the employed fibers are presented in Table 2. The characteristic part of the chemical components of the employed synthetic fibers consists of the polarity groups indicated by the circle in Figure 1. PVA has the highest polarity strength owing to the OH radical, whereas PE has no polarity strength. Ethylene vinyl alcohol (EVOH) consists of the PVA and PE parts, hence its moderate polarity strength between those of PVA and PE fibers. On the other hand, steel cord (SC) is composed of five thin steel fibers of diameter 135 μm, which are twisted together as shown in Figure 2. It also has no chemical polarity. Previous studies by the authors have revealed that bridging fibers with high polarity strengths attract calcium ions in cracks, which facilitates the filling of the cracks by the precipitation of $CaCO_3$ [13,14].

Table 1. Mixture proportions of fiber reinforced cemebtitions composites (FRCCs)

Series	W/B	S/B	SF/B	SP/B	Fiber (vol%)	Number of specimens
SC				–	0.75	4
PE					1.5	4
EVOH	0.45	0.45	0.15	0.009	2.0	3
PVA						3
PE+SC (Hybrid)					0.75+0.75	4

B: binder (cement + silica fume)

Table 2. Properties of employed fibers

Type of Fiber	Density (g/cm^3)	strength (N/mm^2)	Length (mm)	Di
Polyethylene	0.97	2580	6	
Ethylene vinyl alcohol copolymer	1.04	231	5	
Polyvinyl alcohol	1.30	1600	6	

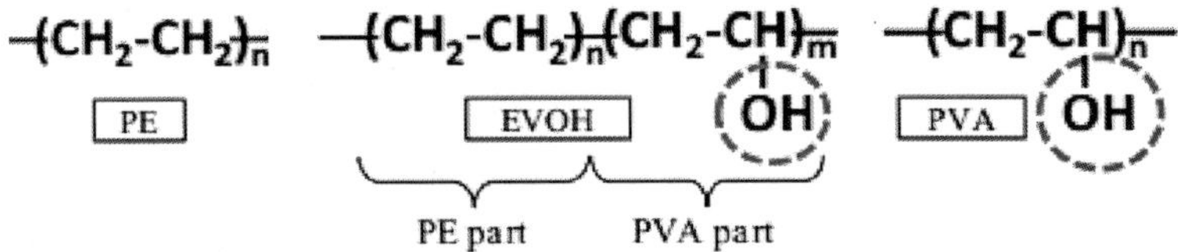

Figure 1. Characteristic part of chemical components.

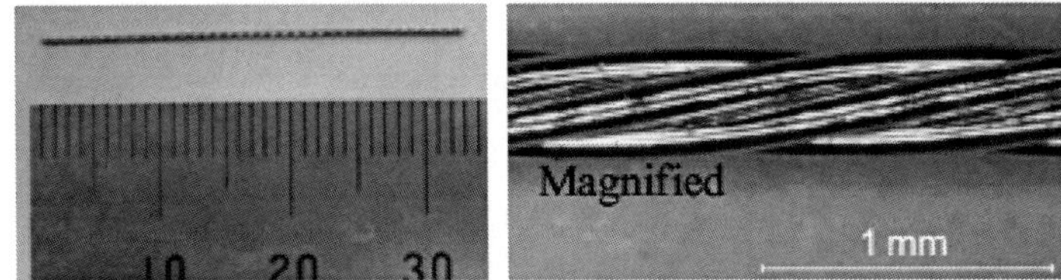

Figure 2. Steel code (SC) fiber.

Preparation of FRCC Specimens

The FRCC mixture was prepared in two liters a Hobart mixer by the following procedure. The cement, silica fume, and sand were first mixed for 1 min. Water and a polycarboxylate-based superplasticizer were then added, and mixing was done for 3 min. Lastly, the fibers were added and mixing was done for another 3 min to achieve proper fiber dispersion for each series. The mixtures were cast in molds. The mold, which measured 85 mm × 85 mm × 30 mm, was prepared with four M6 screw bars, which had anchor nuts at their tips (Figure 3a). The screw bars were embedded in both ends of the specimen to enable holding of the specimen by the testing machine. After de-molding, the specimens were cured in a water tank at 20 °C for 7 days in preparation for testing.

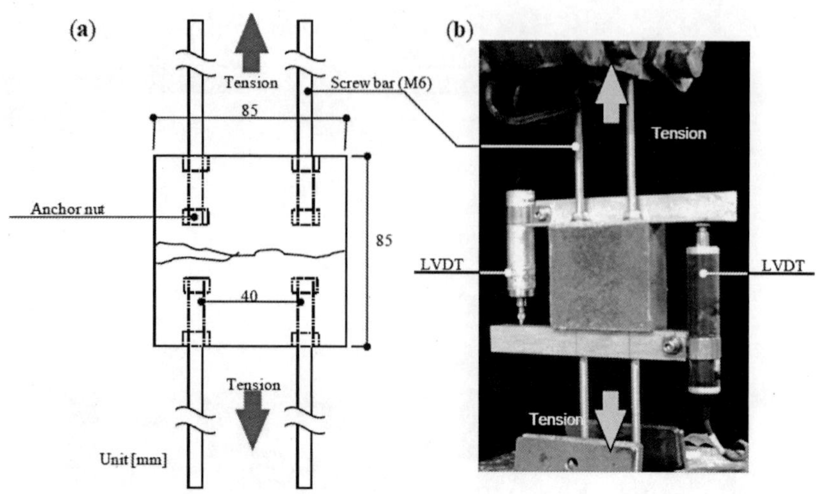

Figure 3. (a) Geometry of specimen and (b) direction of load during tensile test.

Inducement of Cracks during Tensile Loading Test

Direct uniaxial tensile loads were applied to the specimens by a 30 kN capacity load frame under controlled displacement. A quasi-static loading speed of 0.2 mm/min was used. Two external linear variable differential transformers (LVDTs) were attached to the specimen using a gauge length of 85 mm to measure the displacement (Figure 3b).

The experimental procedures are summarized in Figure 4. The specimens were de-molded one day after casting and cured in water for 7 days (first curing). The FRCC specimens were then preloaded to induce three levels of cracks (*i.e.*, <0.1 mm, 0.1–0.5 mm and >0.5 mm) in them.

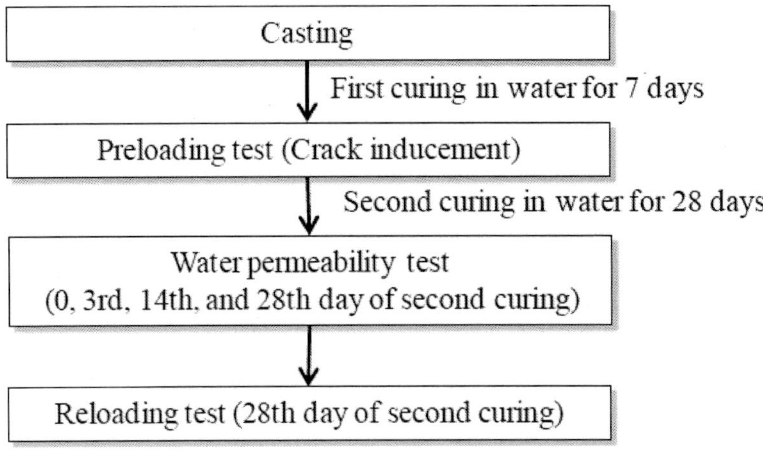

Figure 4. Experimental procedure.

Water Permeability Test and Re-Curing (Second Curing) Method

The first water permeability tests (day 0) were conducted immediately after the tensile test. The apparatus of the water permeability test in this study is shown in Figure 5 [14] (modified after Kishimoto [15]). The coefficient of water permeability was calculated from the volume of water that permeated the specimen. The specimens were cured again in a water tank at 20 °C for 3, 14, and 28 days, respectively (second curing). The water permeability tests were conducted again at the same time after each curing period. After curing for 28 days, the specimens with the induced cracks were reloaded in direct tension test again.

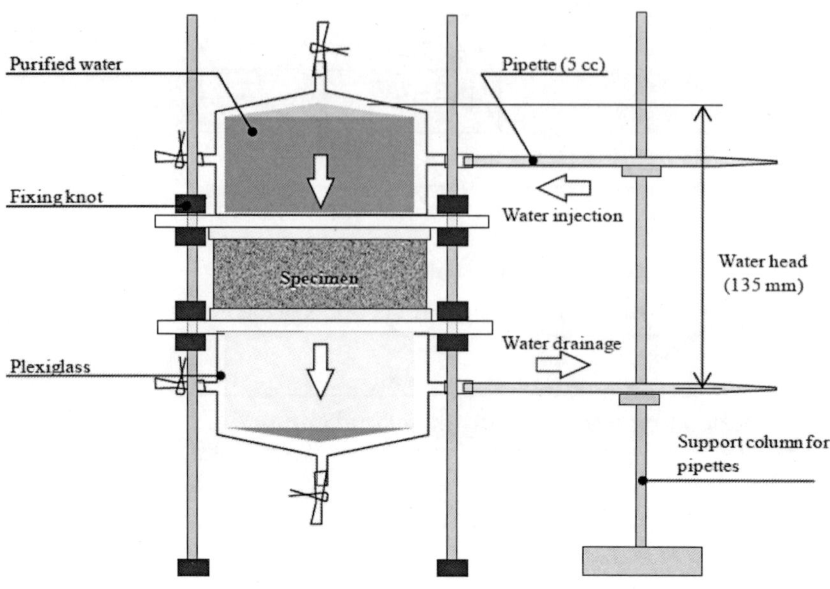

Figure 5. Apparatus of water permeability test.

RESULTS AND DISCUSSION

Recovery of Watertightness

Summarized results of the experimental study are shown in Table 3. This table includes the result of watertightness and mechanical properties which will be discussed in the next section.

Table 3. Summarized results of experimental study.

Series		Max. crack width (mm)	k_i (m/s)	k_{28} (m/s)	R_k (-)	g_1 (N/mm)	g_2 (N/mm)	R_g (%)
SC	1	0.035	1.12×10^{-7}	6.21×10^{-11}	5.53×10^{-4}	1.96	1.71	87.4
	2	0.076	5.63×10^{-6}	1.10×10^{-6}	1.95×10^{-1}	5.01	2.85	57.0
	3	0.088	6.26×10^{-7}	4.23×10^{-9}	6.76×10^{-3}	–	–	–
	4	0.757	2.14×10^{-5}	6.51×10^{-6}	3.04×10^{-1}	2.25	1.24	55.3
PE	1	0.019	2.51×10^{-8}	2.65×10^{-11}	1.06×10^{-3}	1.10	0.76	69.1
	2	0.038	1.26×10^{-7}	9.63×10^{-11}	7.64×10^{-4}	1.18	0.60	51.1
	3	0.119	5.57×10^{-7}	5.84×10^{-9}	1.05×10^{-2}	2.26	0.57	25.3
	4	0.368	1.88×10^{-5}	6.32×10^{-6}	3.36×10^{-1}	–	–	–
EVOH	1	0.287	4.61×10^{-6}	1.33×10^{-6}	2.87×10^{-1}	0.06	1.42×10^{-3}	2.10
	2	0.339	1.44×10^{-5}	1.46×10^{-6}	1.01×10^{-1}	0.06	6.69×10^{-4}	1.14
	3	0.399	1.03×10^{-5}	6.41×10^{-6}	6.25×10^{-1}	–	–	–
PVA	1	0.106	2.28×10^{-8}	2.59×10^{-11}	1.14×10^{-3}	0.55	0.62	112.9
	2	0.185	5.27×10^{-8}	2.17×10^{-11}	4.12×10^{-4}	0.71	0.69	97.4
	3	0.225	2.56×10^{-7}	4.03×10^{-10}	1.57×10^{-3}	1.22	1.06	86.4
PE+SC (Hybrid)	1	0.017	5.35×10^{-8}	1.42×10^{-11}	2.65×10^{-4}	5.32	5.53	103.8
	2	0.081	1.64×10^{-8}	4.44×10^{-12}	2.71×10^{-4}	5.21	5.18	99.4
	3	0.407	6.03×10^{-6}	1.02×10^{-9}	1.69×10^{-4}	5.78	4.85	83.9
	4	0.710	3.84×10^{-6}	9.57×10^{-7}	2.49×10^{-1}	2.28	2.24	98.6

k_i: initial coefficient of water permeability; k_{28}: coefficient of water permeability after 28 days self-healing; R_k: recovery ratio of watertightness; g_1: energy absorption at first loading test; g_2: energy absorption at second loading test and R_g: recovery ratio of energy absorption.

The variation of the coefficient of water permeability as a function of time during the second curing period is shown in Figure 6. The relationships between the maximum crack width and the initial coefficient of water permeability (ki) and between the maximum crack width and the coefficient of water permeability after self-healing (k28) are shown in Figure 7a,b, respectively.

For quantifying effect of self-healing on water permeability the recovery ratio was determined by Equitation (1):

$$R_k = \frac{k_{28}}{k_i}$$

where *R*k is recovery ratio of watertighness; *k*i is the initial coefficient of water permeability and *k*28 is coefficient of water permeability after self-healing. According to this relationship, lower values of *R*k mean large reduction of watertightness of FRCC due to self-healing. The relationship between the maximum crack width and *R*k of all the samples obtained in the present study are shown in Figure 8.

As shown in Figures 6 and 7a, it is confirmed that the wider the crack width, the larger the initial coefficient of water permeability. The recovery of the water permeability was significantly affected by the crack width. The crack widths were measured by a digital microscope. The measured crack widths were divided into three levels, namely, <0.1 mm, 0.1–0.5 mm and >0.5 mm, which are identified by different colors in the legends of Figure 6. As shown in Figure 6, scatters were observed on SC (<0.1 mm) and PE (0.1–0.5 mm) might be caused by difference of the maximum crack width (see Table 3).

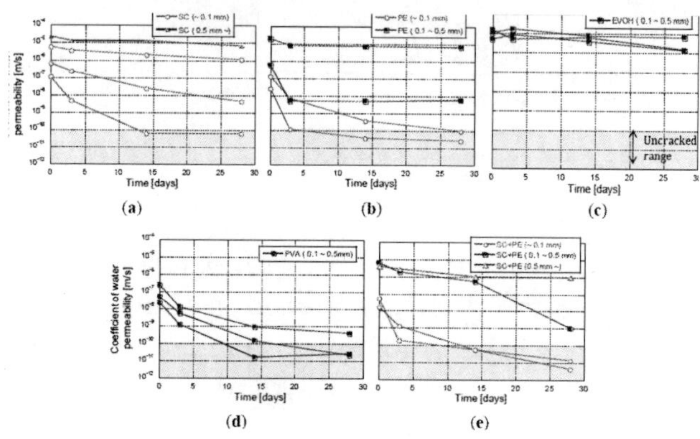

Figure 6. Relationship between time and coefficient of water permeability: (a) SC; (b) polyethylene (PE); (c) ethylene vinyl alcohol (EVOH); (d) polyvinyl alcohol (PVA) and (e) PE+SC.

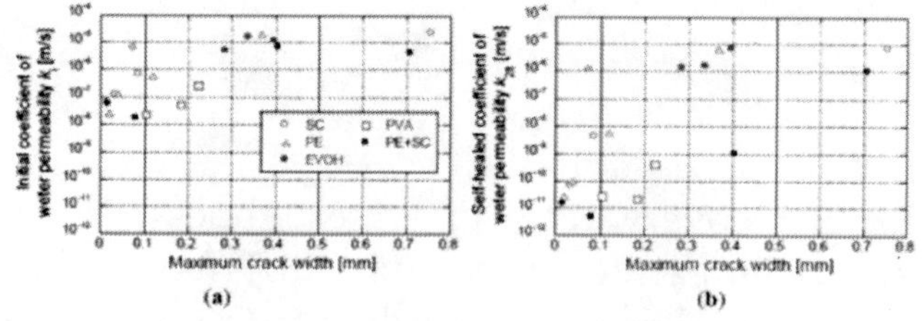

Figure 7. Relationship between the maximum crack width and coefficients of water permeability: (a) initial and (b) after self-healing.

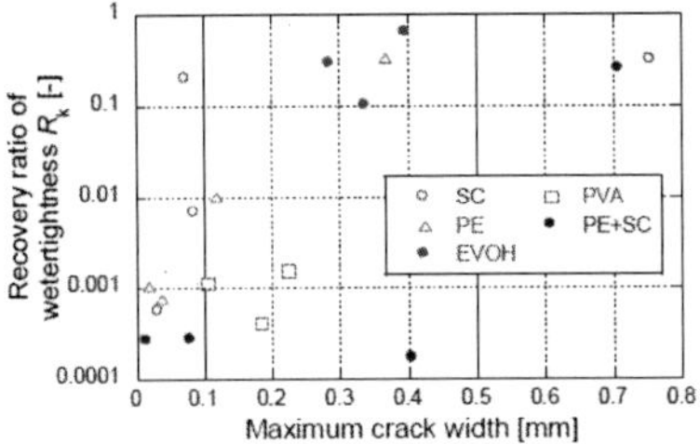

Figure 8. Relationship between the maximum crack width and recovery ratio of watertight ness.

With the exception of two specimens in the SC series, for a crack width less than 0.1 mm, the coefficient of water permeability after the second curing was below 1 × 10−10 (m/s), which was of the same level as that of the uncracked FRCC (blue-hatched area in the graphs). The maximum crack width strongly affects the capability for self-healing of watertightness. The natural healing that is inherent to cementitious materials, owing to the deposition of calcium carbonate, could be exhibited for a sufficiently small crack width [16]. Only the PVA series shows good recovery even in case of over 0.1 mm crack width and approximately 0.2 mm (Figure 7b) from the point of *k*28 view. This recovery was enabled by significant precipitation of calcium carbonate in the cracks due to its high polarity. However, EVOH series did not show any recovery of watertightness, even though EVOH has a moderate polarity which was expected to increase self-healing performance by OH radical as shown in Figure 1. This phenomenon might be related to its brittle behavior which led localization of a crack without multiple cracks after the first crack occurred during the first tensile loading test. Therefore, controlling crack width was very difficult as EVOH series. From the results, polarity of fibers had a potential to improve the self-healing performance, controlling crack width is also very important factors on self-healing capability.

Series of relatively small crack width for exhibited good *R*k (below 0.001) as shown in Figure 8. In this figure, both of PE and SC series show high values of *R*k. However, PE + SC series is confirmed to show a good value of *R*k even at 0.4 mm of the maximum crack width which is a larger crack width than those of PVA series (approximately 0.2 mm). This phenomenon might be synergistic effect of two different fibers (PE and SC) for filling up the crack with $CaCO_3$ [12] controlling crack propagation. Because of the effect, recovery of water tightness could be possible in PE+SC series.

Recovery of Mechanical Properties

After the water permeability tests, the second tensile loading test was conducted to determine the effect of self-healing on the mechanical properties. Typical examples of the relationship between the elongation and the tensile stress are shown in Figure 9. A schematic representations of the relationship between the tensile stress and the elongation during the first and the second tensile loading tests is shown in Figure 10. The recovery ratio of energy absorption (R_g) is defined by the following Equation (2):

$$R_g = \frac{g_2}{g_1} \times 100$$

where : g_1 the energy absorption during the first tensile loading test up to the point of unloading elongation, and : g_2 the energy absorption during the second tensile loading test up to the same unloading elongation point of the first tensile loading test. If R_g is higher than 100%, the absorbed energy of the healed specimen would be greater than that of the initial specimen. This means that the mechanical property is completely recovered. However, if R_g is lower than 100%, the mechanical property is partially recovered or not recovered.

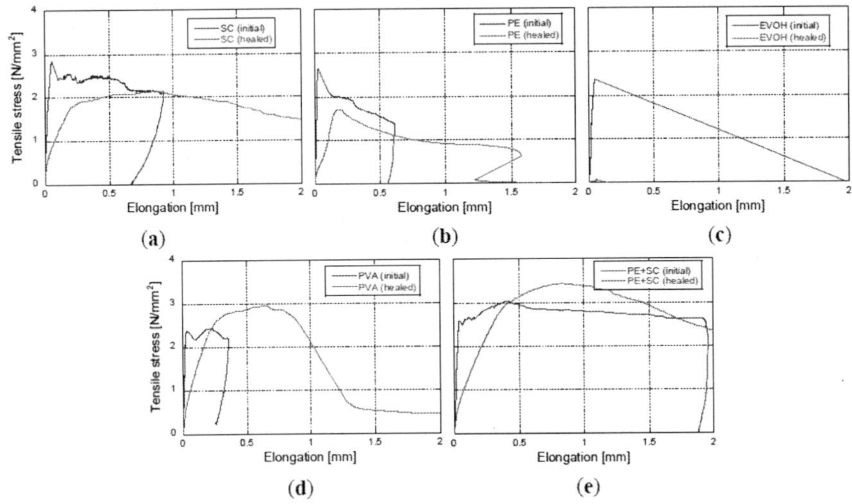

Figure 9. Typical examples of relationship between elongation and tensile stress: SC; (b) PE; (c) EVOH; (d) PVA and (e) PE+SC.

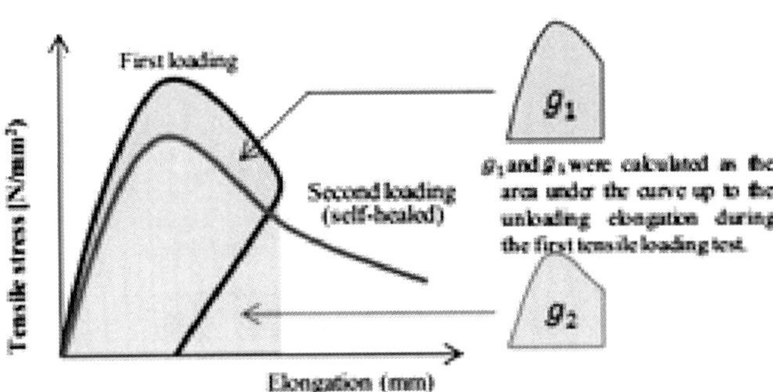

Figure 10. Schematic diagram of relationship between tensile stress and elongation.

The relationship between the recovery ratio of energy absorption R_g and the maximum crack width is shown in Figure 11. For sufficiently small maximum crack width (<0.1 mm), the recovery ratio was relatively high. However, the recovery ratio of SC, PE and

EVOH series did not exceed 90% of . R_g On the other hand, PE+SC and PVA series showed more than approximately 90% and some specimens show superior to 100% of R_g. Those series occurred in case of the fine width of multiple cracks which led to pseudo strain-hardening behavior. The small crack width might be considered as one of the essential properties in order to recover the mechanical properties. Especially, PE+SC series showed almost 100% of R_g at maximum crack width approximately 0.7 mm. $CaCO_3$ fills and closes up the micro cracks and bond strength of SC works for the recovery. Basically, the self-healing mechanism of FRCCs is precipitation of $CaCO_3$ and it does not contribute to regain mechanical property [16]. Yang *et al.* [17] was confirmed the mechanical recovery of PVA/ matrix by reaction of unhydrated cement and/or fly ash. In this study, although relative higher water binder ratio than their matrix is applied, unhydrated cement still exists in the matrix, which is confirmed by Homma *et al.* [12]. In addition, a pozzolantic reaction of silica fume also expected to positive effect to mechanical recovery.

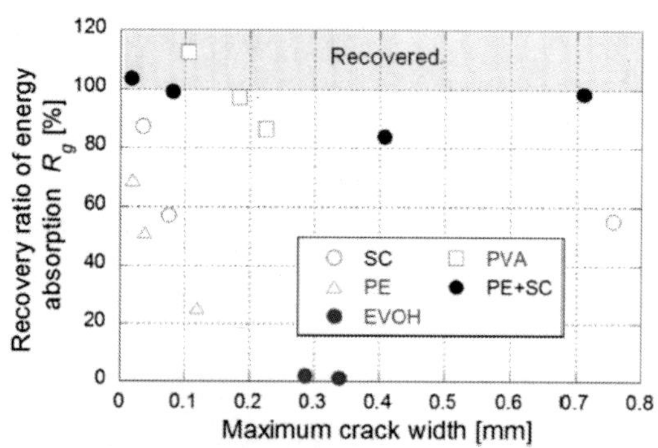

Figure 11. Relationship between recovery ratio of energy absorption ( ) and maximum crack width

Regarding the mechanical recovery of PE+SC series, chemical bond of each fiber is not expected. Therefore, PE+SC series might have different self-healing mechanism from that of PVA series. In addition, PE+SC series is able to be recovered its mechanical property almost up to 100%, even in case that macro crack is occurred (approximately

0.7 mm). This mechanism might be explained by pullout procedure as shown in Figure 12, which illustrates a schematic diagram of the pullout behavior of different types of fibers in the matrix. The PE fiber (microfiber) can control very fine cracks by the bridging effect and snubbing effect [18] (Figure 12a). Whereas SC fiber (macro fiber) is also not expected to form a chemical bond, it forms a strong mechanical bond owing to its unique uneven geometry (Figure 2). An SC fiber can resist visible crack propagation of very fine cracks (Figure 12b). Based on the reinforcement mechanism of a hybrid fiber reinforcing system, that is, long and thick macrofibers and short and thin microfibers are blended, it can be possible to be expected a synergistic effect [19]. By this mechanism, it is expected that the PE+SC series would be able to control both of very fine cracks and visible macro cracks. The PE fiber would support the reinforcement of fine cracks around the SC fiber. On the other hand, the stiff SC fiber can resist macro cracks. Thus, the widths of the cracks are restricted to small values (Figure 12c). This phenomenon was confirmed by the following single fiber pullout test. A schematic diagram of the specimen used for the single fiber pullout test is shown in Figure 13. Fiber pullout specimens with SC fiber inclined at 0 degree were prepared. SC fibers were embedded up to 16 mm (half of the total length) in two different matrices, namely plain matrix (PM) of the same mix proportion shown in Table 1 and matrix reinforced with PE (volume fraction = 0.75%) (PEM). The size of the pullout specimen was 40 mm × 40 mm × 80 mm. After the pullout test, spalling of the matrix was observed (Figure 14). Cone-type failure of the PM was observed at the fiber exit point (Figure 14a). Conversely, very small cracks without spalling were observed around the fiber exit point in the PEM (Figure 14b). These small cracks were filled and closed by self-healing, which was accelerated by the synergistic effect due to PE fiber. In other words, in the case of PE+SC series, small size of cracks around SC fiber was filled by hydration products of unhydrated cement particles, silica fume which activated the pozzolanic reaction, and precipitation of $CaCO_3$. Therefore matrix might be densified and performance of mechanical bond was also recovered. From this point of view, mechanical recovery at 0.7 mm of PE+SC series was also obtained by recovery of mechanical bond of SC even though macrocracks were not healed. Self-healing mechanism by the ongoing hydration and aforementioned mechanism of PE+SC series, further experimental researches are required.

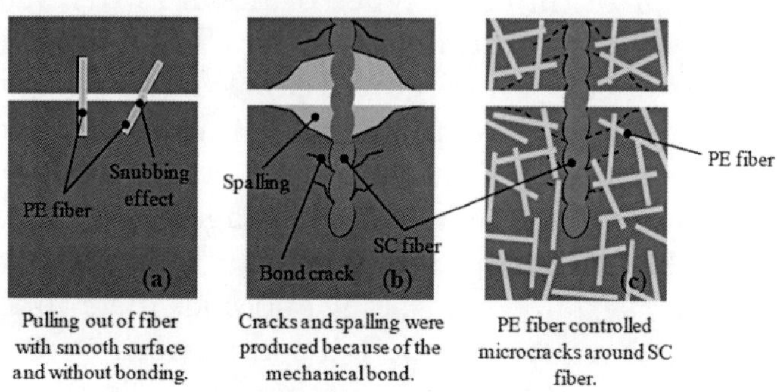

Figure 12. Schematic diagram of pullout behavior of different types of fibers in the matrix. (a) Straight fiber without bonding; (b) SC fiber and (c) SC fiber in FRCC matrix.

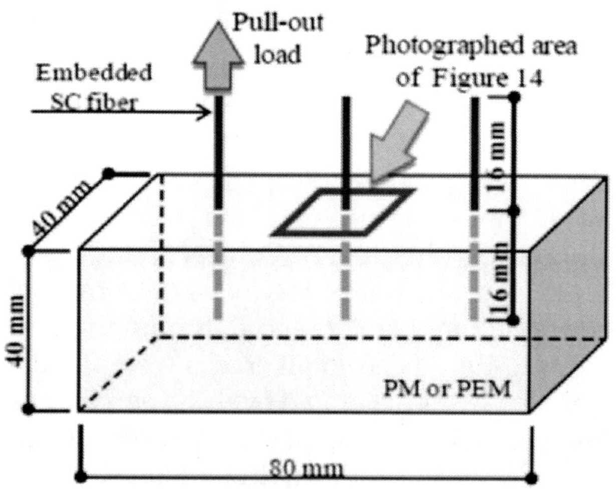

Figure 13. Specimen used for single fiber pullout test.

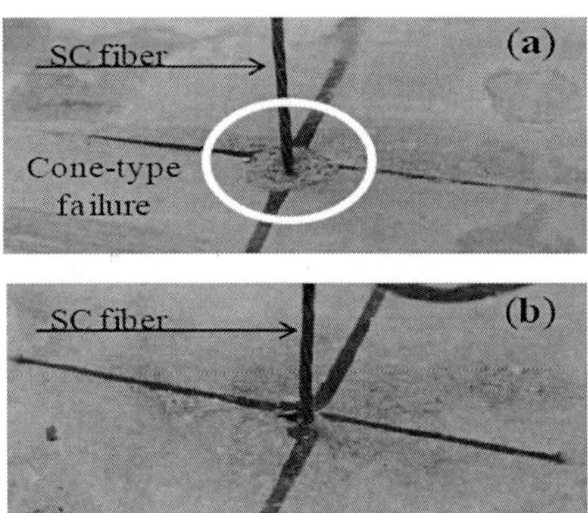

Figure 14. Matrix failure around fiber exit point: (a) PM and (b) PEM.

CONCLUSIONS

In this study, the self-healing capability of FRCCs was investigated using different types of fibers. Based on the experimental results, the following conclusions are drawn.

- If the crack in a FRCC containing synthetic fibers is narrower than 0.1 mm, the watertightness would be recovered regardless the type of fiber as same as watertightness of uncracked FRCC. PVA series, which has polarity, exhibits recovery of the watertightness for crack width of up to approximately 0.2 mm. However, EVOH series, which also has polarity, did not show significant recovered watertightness since the crack width was not well controlled because of the brittle behavior of FRCC. PE+SC series exhibits relatively high recovery ratio of water tightness even in case of the maximum crack widths of 0.4 mm. This is because it generated multiple cracks with fine crack width on the specimens. Recovery ratio of watertightness significantly depends on the crack width. Therefore, controlling crack width is one of important factors in self-healing.
- Regarding the mechanical properties of FRCCs after self-healing, a very high recover ratio was confirmed in PVA and PE+SC series. In particular, PE+SC series recovered their mechanical property up to approximately 100% recover ratio

even when the crack width was approximately 0.7 mm. This might be because bond crack in the matrix around SC fiber was bridged by PE fiber and the crack width was well controlled. Thus, bond cracks were densifed by self-healing and it linked to the mechanical recovery.

ACKNOWLEDGMENTS

This research was partially supported by a Grant-in-Aid for Young Scientists (A) from the Ministry of Education, Culture, Sports, Science and Technology of Japan (#23686078, 2011-2014).

AUTHOR CONTRIBUTIONS

In this study, Daisuke Homma has performed experimental works regarding recovery of mechanical properties of PE, SC and SC+PE series under supervision of Hirozo Mihashi. Sukmin Kwon and Makoto Yamada have performed experimental works regarding recovery of mechanical properties of PVA and EVOH series under supervision of Tomoya Nishiwaki. Sukmin Kwon has also performed single fiber pullout test regarding the synergistic effect of SC+PE series. The manuscript was written and edited by Tomoya Nishiwaki via scientific discussion with Hirozo Mihashi and Sukmin Kwon.

REFERENCES

1. 1. Mihashi, H.; Nishiwaki, T. Development of engineered self-healing and self-repairing concrete—State-of-the-art report. *J. Adv. Concr. Technol.* **2012**, *10*, 170–184.
2. 2. Van Tittelboom, K.; De Belie, N. Self-healing in cementitious materials—A review. *Materials* **2013**, *6*, 2182–2217.
3. 3. De Rooij, M.; Van Tittelboom, K.; De Belie, N.; Schlangen, E. *Self-Healing Phenomena in Cement-Based Materials, State-of-the-Art Report of RILEM Technical Committee 221-SHC: Self-Healing Phenomena in Cement-Based Materials*; Springer: Dordrecht, the Netherlands, 2013.
4. 4. Jonkers, H.M. Self Healing Concrete: A Biological Approach. In *Self Healing Materials: An Alternative Approach to 20 Centuries of Materials Science*; Van Der Zwaag, S., Ed.; Springer: Dordrecht, the Netherlands, 2007; pp. 195–204.
5. 5. Wiktor, V.; Jonkers, H.M. Quantification of crack-healing in novel bacteria-based self-healing concrete. *Cem. Concr. Compos.* **2011**, *33*, 763–770.

6. 6. Nishiwaki, T.; Mihashi, H.; Jang, B.K.; Miura, K. Development of self-healing system for concrete with selective heating around crack. *J. Adv. Concr. Technol.* **2006**, *4*, 267–275.

7. 7. Nishiwaki, T.; Mihashi, H.; Okuhara Y. Fundamental Study on Self-Repairing Concrete Using a Selective Heating Device, In *Proceedings of the Sixth International Conference on Concrete under Severe Conditions: Environment & Loading* (*CONSEC'10*), Mérida, Mexico, 7–9 June 2010; Castro-Borges, P., Morenoe, E., Sakai, K., Gjorv, O., Banthia, N., Eds., Taylor & Francis Ltd.: Boca Raton, FL, USA, 2010; Volume 2, pp. 919–926.

8. 8. Dry, C.M. Three designs for the internal release of sealants, adhesives, and waterproofing chemicals into concrete to reduce permeability. *Cem. Concr. Res.* **2000**, *30*, 1969–1977.

9. 9. Van Tittelboom, K.; De Belie, N.; Van Loo, D.; Jacobs, P. Self-healing efficiency of cementitious materials containing tubular capsules filled with healing agent. *Cem. Concr. Compos.* **2011**, *33*, 497–505.

10. 10. Joseph, C.; Jefferson, A.D.; Isaacs, B.; Lark, R.J.; Gardner, D.R. Adhesive Based Self-Healing of Cementitious Materials. In Proceedings of the 2nd International Conference on Self Healing Materials, Chicago, IL, USA, 28 June–1 July 2009.

11. 11. Pareek, S.; Oohira, A. A Fundamental Study on Regain of Flexural Strength of Mortars by Using a Self-repair Network System. In Proceedings of the 3rd International Conference on Self Healing Materials, Bath, UK, 27–29 June 2011.

12. 12. Homma, D.; Mihashi, H.; Nishiwaki, T. Self-healing capability of fibre reinforced cementitious composites. *J. Adv. Concr. Technol.* **2009**, *7*, 217–228.

13. 13. Koda, M.; Mihashi, H.; Nishiwaki, T.; Kikuta, T.; Kwon, S.M. Self-Healing Capability of Fiber Reinforced Cementitious Composites, In *Proceedings of the 3rd International Symposium on Advances in Concrete through Science and Engineering*, Hong Kong, China, 5–7 September 2011; Leung, C., Ed.; International Union of Laboratories and Experts in Construction Materials, Systems and Structures (RILEM): Hong Kong, China, 2011, pp. 543–551.

14. 14. Nishiwaki, T.; Koda, M.; Yamada, M.; Mihashi, H.; Kikuta, T. Experimental study on self-healing capability of FRCC using different types of synthetic fibers. *J. Adv. Concr. Technol.* **2012**, *10*, 195–206.

15. 15. Kishimoto, Y.; Hokoi, S.; Harada, K.; Takada, S. The effect of vertical distribution of water permeability on the modeled neutralization process in concrete walls. *J. ASTM Int.* **2007**, *4*, doi:10.1520/JAI100323.

16. 16. Edvardsen, C. Water permeability and autogenous healing of cracks in concrete. *ACI Mater. J.* **1999**, *96*, 448–454.

17. 17. Yang, Y.; Lepech M.D.; Yang, E.H.; Li, V.C. Autogenous healing of engineered cementitious composites under wet-dry cycles, *Cem. Concr. Res.* **2009**, *39*, 382–390.

18. 18. Wu, H.C.; Li, V.C. Snubbing and bundling effects on multiple crack spacing of discontinuous random fiber-reinforced brittle matrix composites.

J. Am. Ceram. Soc. **1992**, *75*, 3487–3489.

19. 19. Kwon, S.; Nishiwaki, T.; Kikuta, T.; Mihashi, H. Tensile behavior of ultra high performance hybrid fiber reinforced cement-based composites. In *Proceedings of 8th International Conference on Fracture Mechanics of Concrete and Concrete Structures* (*FraMCoS-8*); Toledo, Spain, 11–14 March 2013; Van Mier, J.G.M., Ruiz, G., Andrade, C., Yu, R.C., Zhang, X.X., Eds.; CIMNE: Barcelona, Spain, 2013; pp. 1309–1314.

Chapter 12

NONLINEAR ANALYSES OF ADOBE MASONRY WALLS REINFORCED WITH FIBERGLASS MESH

Vincenzo Giamundo, Gian Piero Lignola *, Andrea Prota and Gaetano Manfredi

Department of Structures (Di.St.), University of Naples Federico II, Via Claudio 21, Naples 80125; Italy

ABSTRACT

Adobe constructions were widespread in the ancient world, and earth was one of the most used construction materials in ancient times. Therefore, the preservation of adobe structures, especially against seismic events, is nowadays an important structural issue. Previous experimental tests have shown that the ratio between mortar and brick mechanical properties (i.e., strength, stiffness and elastic modulus) influences the global response of the walls in terms of strength and ductility. Accurate analyses are presented in both the case of unreinforced and reinforced with fiberglass mesh when varying the mechanical properties of the materials composing the adobe masonry structure. The main issues and variability in the behavior of seismic resisting walls when varying the mechanical properties are herein highlighted. The aim of the overall research activity is to improve the knowledge about the structural behavior of adobe structural members unreinforced and reinforced with fiberglass mesh inside horizontal mortar joints.

INTRODUCTION

Earth is one of the oldest and most widespread construction material used in the construction of buildings and has been used for thousands of years. Earth can be considered the cheapest material readily available for construction. It has been extensively used for masonry constructions around the world. The popularity of adobe is also related to the low level of skill and technology required for the production of the bricks and construction. Furthermore, due to its inherent properties, earth is an efficient heat and sound insulating material [1,2]. After the introduction of reinforced concrete, the number of adobe buildings has become progressively less. Nevertheless, as reported in [3], it is estimated that approximately 30% of the world's population still lives in buildings made of earth. In spite of its past and present spread, adobe constructions are prone to damage under seismic actions [4], and most of these buildings are located in high seismic risk areas. Therefore, there is nowadays raising awareness for the preservation of historic, archeological, or even still in use structures. Unfortunately research on the structural behavior of adobe buildings, and especially on their reinforcement, is inadequate, so far. The seismic resistance of adobe masonry structures primarily depends on the in-plane shear behavior of individual walls. The typical failure mode under earthquakes is the diagonal cracking mode. This occurs when the principal tensile stresses developed in the wall, under a combination of vertical and horizontal loads, exceed the tensile strength of the adobe material. The seismic characterization of the masonry structural element is generally carried out through direct shear tests or diagonal compression tests. According to [5], the diagonal compression test allows the determination of the diagonal tensile (or shear strength) of masonry assemblages. The test consists in loading the specimen in compression along one diagonal, resulting in a diagonal tension failure. This test allows for the evaluation of the effects of variables, such as the type of masonry unit, mortar, workmanship, etc. [6]. The use of plaster reinforcement mesh between adobe blocks, in the construction phase, can significantly improve the shear strength of adobe walls. Fiberglass mesh is a glass fiber fabric, usually coated with a polymer latex-resistant matrix. It is widely used in the building industry as a plaster stiffener or as wall reinforcement (anti-crack). Previous experimental tests by [7] showed the effect of fiberglass reinforcement on adobe masonry walls. These tests represent the base of the present paper.

SUMMARY OF EXPERIMENTAL OUTCOMES

Test Setup

Experimental diagonal compression tests on adobe wall panels (unreinforced and reinforced with plaster fiberglass mesh) presented in [7] have been used as a benchmark for the validation of the herein presented numerical analyses. All the tested panels were subjected to diagonal compressive load (compressive edge load) acting in the plane of the wall and forming a 45° angle with the direction of the mortar bed joints; the geometry of a tested panel is reproduced in Figure 1.

The specimens were built with the global size of 80 × 80 cm2. The brick unit size was 11.5 × 10.5 × 21.5 cm3. The mortar thickness was 10 mm, while the width (the third out-of-plane dimension) was 11.5 cm for both bricks and mortar. Different adobe soil curing for both the bricks and the mortar were used for each specimen. The reinforced panels were built placing a plaster reinforcement fiberglass mesh (with equal spacing of 5 mm × 5 mm) inside the horizontal mortar joint of the adobe blocks. This sort of mesh fiber reinforcement is very cheap and commonly used for reinforced plaster coatings, usually applied on the exterior and interior faces of walls by the construction industry.

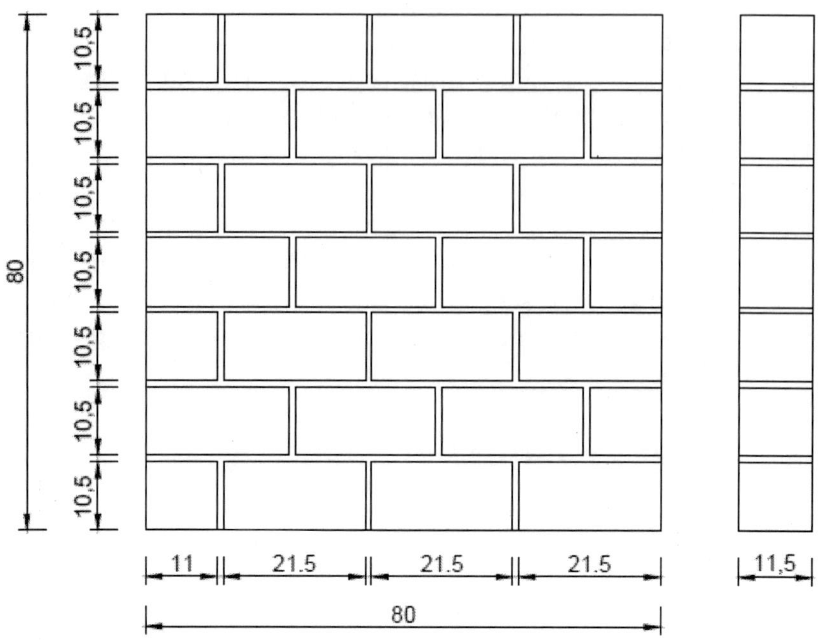

Figure 1 Tested panel geometry.

Material Characterization

An essential mechanical characterization of the material is provided in [7]. Nevertheless, further data not present in [7] have been achieved by fitting numerical global experimental data. Tensile and compressive strengths of both the brick and the mortar, at different curing times, were achieved by interpolation according to the values suggested in [8]. Such tests provide data from 0 to 28 days of curing for plain soil. The elastic modulus, E, of the adobe soil was achieved by fitting the global response curve, because such mechanical characterization was lacking. According to the tests presented in [8], the trends of the compressive and tensile strength ratio when varying the curing time are shown in Figure 2. The strength ratio is defined as the ratio between the actual strength (related to the curing time) and the maximum strength. As shown in Figure 2 the compressive strength of the plain soil is almost doubled after about three weeks. Being that both the mortar and the bricks made of the adobe soil, the increase of strength should be considered either for brick and mortar according to their curing time.

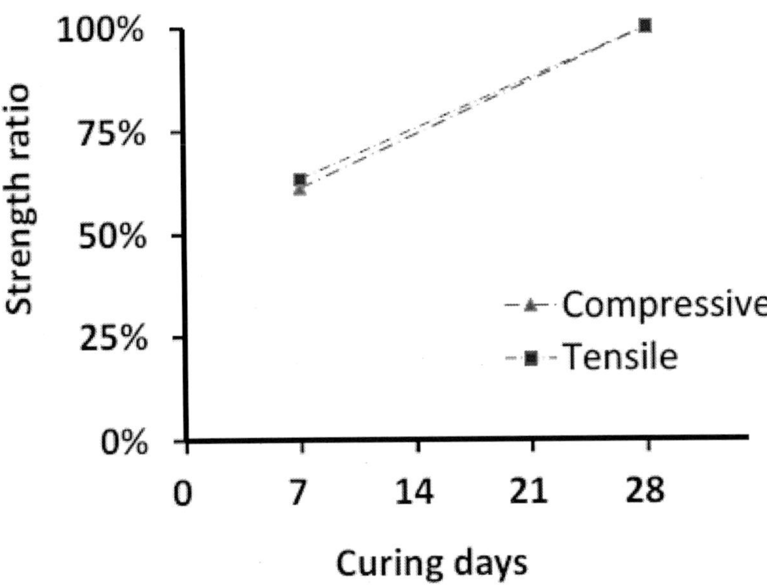

Figure 2 Compressive and tensile strength trends when varying the curing time.

NONLINEAR NUMERICAL ANALYSES

Finite Element Method

The influence of the different constituent materials (*i.e.*, the local properties of soil, composition, curing), as well as the fiberglass plaster mesh reinforcement on the global structural behavior has been studied by means of a finite element method (FEM) model. The FEM model has been validated comparing the numerical FEM outcomes to the experimental data [7]. Micro modeling was adopted and some parametric analyses were conducted on validated models.

Numerical two-dimensional analyses have been performed. Since the width of the panel is much lower than the other two dimensions, a plane-stress assumption was adopted, and in-plane loads were applied. Different mechanical properties are provided for plain adobe soil when varying the curing time. In particular, the analyses were performed when varying the strength (tensile and compressive) of both, the bricks and the mortar, according to the curing time. In particular, bricks at the testing time [7] were 28 days old, while mortar was almost fresh. Typical values for the fiberglass mesh mechanical parameters have been used.

In particular, according to the experimental tests, a fiberglass mesh with the following properties has been considered: weight 100 g/m2, density 2.5 g/cm3, elastic modulus 20 GPa and strength 1000 N/50 mm. Lacking an experimental counterpart, for comparison purposes, numerical simulations were performed considering two more cases: the first one applying a mortar having higher performances compared to plain fresh mortar, namely, "mortar B"; the second one considering for both mortar and bricks the same properties, due to the long time of the curing (*i.e.*, longer than 28 days), namely "long-term curing". The adopted values for the tensile and compressive strength, according to curing time, are reported in Table 1.

Table 1 Material mechanical properties.

Properties	Fresh plain soil	Plain soil	Plain soil	Mortar B
Curing time (days)	0	28	>28	0
Tensile strength (kPa)	39.2	78.4	98	98
Compressive strength (MPa)	3.32	6.64	8.30	8.30

The FEM model used is constituted of more than 13,000 eight-node quadrilateral isoperimetric plane stress elements based on quadratic interpolation and Gauss integration (Figure 3b).

All the analyses have been performed by means of the TNO DIANA v9.4.4 code. In the FEM model, bricks and mortar are modeled individually, without interface elements between them, according to the total strain model coupled with the rotating crack stress-strain relationship approach. In particular, in the total strain approach, the constitutive model describes the stress as a function of the strain. In the rotating crack approach, stress-strain relationships are evaluated in the principal directions of the strain vector, as reported in [9]. Furthermore, the combined Rankine/von Mises yield criterion was adopted. Interface elements have been neglected, according to previous studies [6,10–12], mainly due to the lack of experimental properties. The reinforcement has been modeled as bar reinforcements embedded in all plane stress elements at the horizontal mortar joint locations. In particular, 1D bar reinforcement elements have been used; for these elements, the strains are computed from the displacement field of the mother elements. Thus, a perfect bond between the reinforcement and the surrounding material is assumed [9]. The total area of the reinforcement grid cross-section has been modeled as the equivalent area concentrated in the reinforcement bar cross-section.

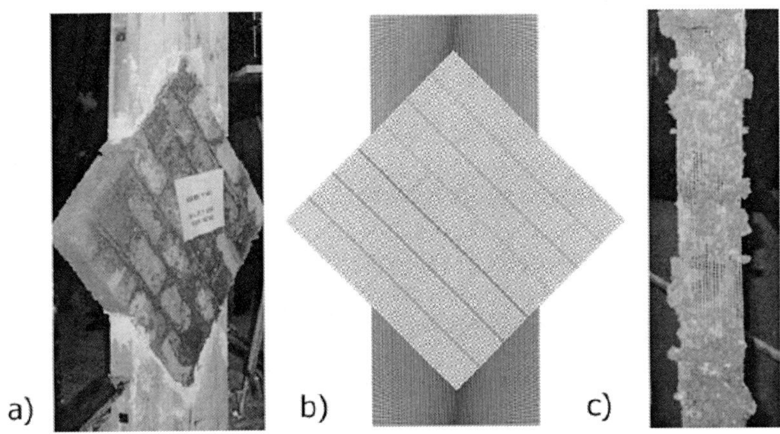

Figure 3 Test setup: (**a**) experimental; (**b**) finite element model; and (**c**) removed fiberglass mesh from the horizontal mortar joint ((**a**) and (**c**) were reprinted from [7], Copyright 2001 Elsevier).

Few data were available for constituent materials, especially for the nonlinear post peak phase; therefore, ideal plasticity was assumed in compression. This assumption is acceptable, since the compressive strength has never been reached during the analyses, remarking that the tensile behavior governs the problem. Therefore, in tension, two limit cases were considered, namely "ideal plasticity" and "brittle failure". The "brittle failure" represents the worst case, which is the more realistic, as well. This case has been modeled by means of an elastic-brittle model. Whereas the "ideal plasticity", which represents the upper bound, has been modeled by means of an elastic-perfectly plastic model. These two cases have been considered in order to define the boundaries between which the real behavior has to be. All the analyses were performed under displacement control, measuring in-plane deformations and the evolution of reacting stresses. The diagonal compressive axial load has been applied, as a displacement load, through two wooden supports. The supports have been modeled at the two opposite corners of the panel, according to the experimental test setup [7]. Eight-node quadrilateral isoperimetric plane stress elements were used to model the supports, as well. As a boundary condition, the bases were fixed.

Numerical Program

The results of FEM analyses were validated through a comparison between experimental and numerical outcomes. In particular, three main cases were considered, each one including both the tensile plasticity models introduced in the previous section (i.e., ideal plasticity and brittle failure.).

Moreover, for each main case, both reinforced and unreinforced wall configurations have been analyzed. The three main cases considered have been named as follows:

- Plain tested: the wall is made of plain adobe bricks, 28 days curing and fresh plain mortar (i.e., 0 day curing), the elastic modulus for both materials is 20 MPa based on experimental data fitting;
- Long-term curing (LTC): this wall, not tested in reality, is made of plain adobe bricks and plain mortar after a long curing time (i.e., a curing time higher than about 28 days); elastic moduli were the same as the first case, for comparison purposes;
- Mortar B (MB): this wall, not tested in reality, is considered only for comparison purposes with the plain tested wall; the wall is made of plain adobe bricks, 28 days curing and a better mortar compared to plain soil; elastic moduli were the same as the first case, for comparison purposes.

The masonry material properties used in the FEM analyses are listed in the following Table 2. The plain soil properties are based on experimental outcomes [7,8]. Conversely, since LTC and MB panels were not tested in reality, their properties are based on the material property trends listed in Table 1. The elastic modulus is derived from best fitting of the global experimental curve for a plain unreinforced panel.

Table 2 Material parameters used for the finite element method (FEM) analyses. LTC, long-term curing; MB, Mortar B.

FEM model	Brick tensile strength (kPa)	Mortar tensile strength (kPa)	Brick compressive strength (MPa)	Mortar compressive strength (MPa)
Plain	78.4	39.2	6.64	3.32
LTC	98.0	98.0	8.30	8.30
MB	78.4	98.0	6.64	8.3

OUTCOMES OF NUMERICAL ANALYSES

Experimental Theoretical Comparison

The behavior of adobe wall panels (unreinforced and reinforced with fiberglass mesh) was analyzed (when varying tension softening models) in terms of the force/displacement curve, shear stress-average diagonal strain curve, shear stress-average shear strain curve, Poisson ratio-displacement and Shear modulus-displacement curves. According to [5], for the standard method, the shear stress, τ, has been computed as $\tau = 0.707 \cdot V/A_n$, where V = diagonal load and A_n = the net section area of the un-cracked section of the panel (in the considered case, $A_n = 0.092$ m2). The average vertical and horizontal strains, ε_v and ε_h, have been computed as the average displacement along the compressive and tensile diagonals, respectively, over the same gauge length (400 mm). The shear strain, γ, according to [5], is:

$\gamma = \varepsilon_v + \varepsilon_h$. The shear modulus, G, and the Poisson ratio, v, were computed according to the well-known solid mechanics relationships, as $v = -\varepsilon_h/\varepsilon_v$ and $G = \tau/\gamma$, respectively. Failure modes, in all the considered cases, were also checked by means of the crack patterns. The experimental failure mode mainly involved the cracking and detachment of lateral corners of the panel, outside the compressed strut between the two wooden loading supports. In the considered cases, the numerical outcomes mainly showed the same cracking pattern, but the spreading of cracks (smeared crack strain field in numerical simulations) depends mainly on the post peak tensile behavior of the soil. In the next figures, solid signs represent experimental data, while continuous and dashed lines represent brittle and ideal post-peak tensile behavior, respectively (i.e., the adopted tensile plasticity model), in FEM outcomes. A small cross marks the failure point for brittle material.

FEM Models: Validation

The FEM model has been validated both in the cases of unreinforced and reinforced plain panels. In both the cases, mortar and bricks are made of the same material. However, different curing times differentiate the two materials. In particular, the mortar is almost fresh, while bricks are cured for 28 days. Therefore, the strength of the brick is higher in both tension and compression. The main outcomes of the numerical analyses and a comparison between the numerical and experimental force/displacement curves are presented in Figure 4 in the case of an unreinforced panel.

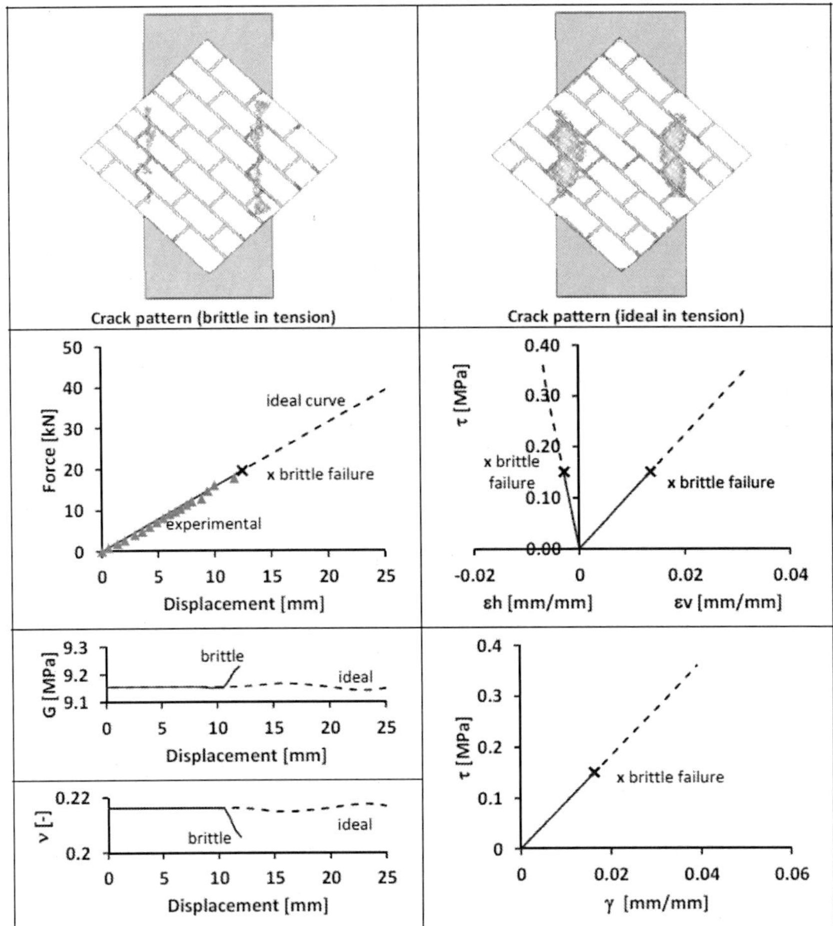

Figure 4 Experimental-theoretical comparison (unreinforced plain adobe soil).

According to the results shown in Figure 4, the brittle material model seems to catch the experimental failure better, in terms of both failure load and the crack pattern. Actually, as shown in Figure 5, the crack pattern is widely smeared, and it almost involves vertical lines connecting the wooden supports.

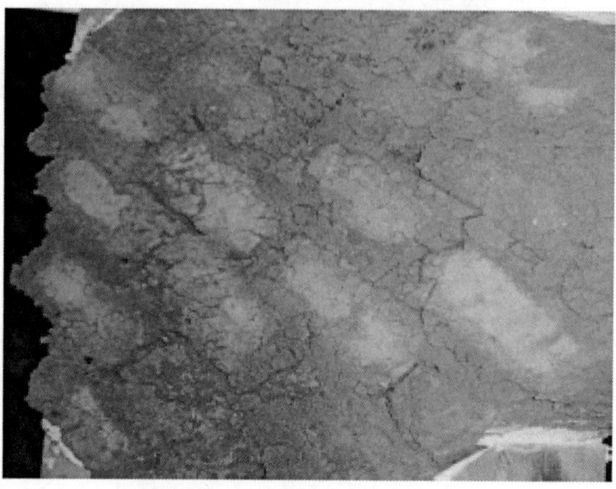

Figure 5 Detail of the experimental crack pattern (unreinforced plain adobe soil) (reprinted with permission from [7], Copyright 2001 Elsevier).

The global response of the reinforced panel in terms of force/displacement highlights, as well as the unreinforced panel is an almost linear behavior (see Figure 6).

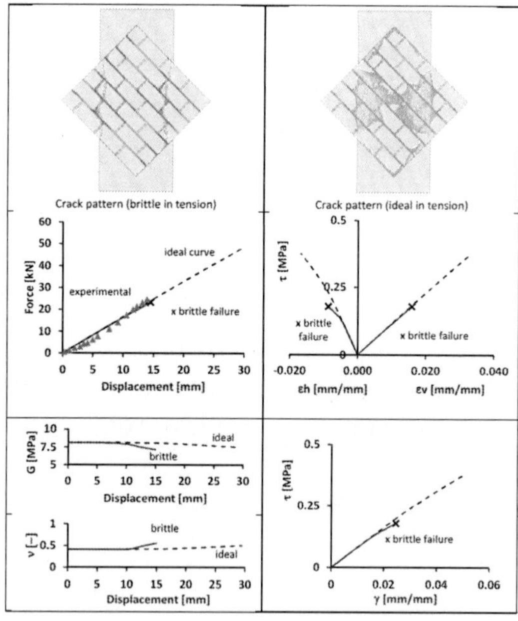

Figure 6 Experimental-theoretical comparison (reinforced plain soil panel).

The brittle material model catches the experimental failure better, both in terms of crack pattern and failure load. In the case of the brittle material, the shear modulus, G, as well as the Poisson ratio, ν, exhibits, up to the failure, the same trend as the case of the ideal material. Of course, the ideal material yields to a longer loading branch. The strength of the plain panel with mesh is about 25% higher, compared to plain soil without mesh, and the ultimate displacement is about 50% higher, so that the benefits of a thin fiber mesh are evident.

FEM Models: Long-Term Curing (LTC)

The LTC panels were not experimentally tested, but they are analyzed to assess the influence of curing time (*i.e.*, strength) of materials on the global behavior of the masonry panels. Since mortar and bricks have identical mechanical properties, these panels can be considered as practically homogeneous panels. The strengths of the two materials correspond to more than 28 days curing, and it is about 25% higher, compared to plain soil bricks (28 days curing) and 2.5 times higher than plain soil mortar (0 day curing). The global response in terms of force/displacement is almost linear up to the failure point (Figure 7). The shear strength achieved is almost doubled if compared to the first unreinforced plain panel. In any case, the main difference, compared to the first test, is the strength of the materials. The global effect is an increase of the shear strength, being comparable to the increase of the mortar strength. The crack pattern of the unreinforced panel is almost restricted to the vertical lines connecting the wooden supports, but it is quite smeared, due to the higher strength of this panel.

In the case of reinforced LTC panel (Figure 8), the crack pattern achieved is similar to the one achieved in the case of the reinforced plain panel (Figure 6). However, a wider crack pattern was found, due to the higher mechanical properties. The global response in terms of force/displacement is almost linear up to the failure point in the case of the ideal material. However, in the case of brittle material, the graph has a second, less stiff branch after a displacement of 17 mm. The stiffness reduction is related to global cracking damage coupled with a shear modulus, G, drop and the Poisson ratio, ν. The difference, compared to the plain panel with mesh, is the strength of materials, and the global effect is an increase of the shear strength comparable to the increase of the basic material strength (about doubled). Compared to the plain panel without mesh, an increase of strength of about 2.3 times is noticed, while the ultimate displacement is almost doubled.

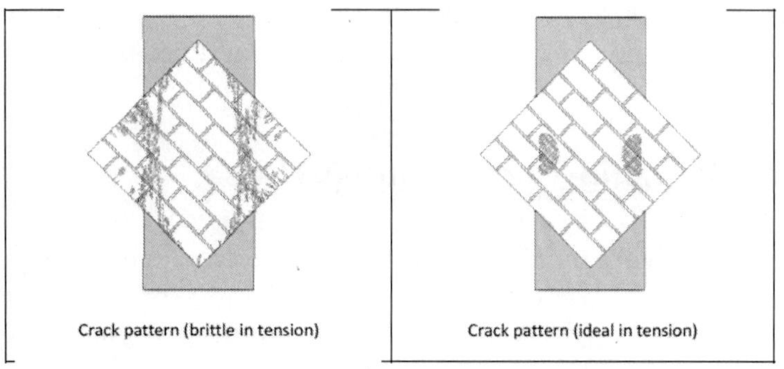

Figure 7 Numerical experimentation (unreinforced LTC panel).

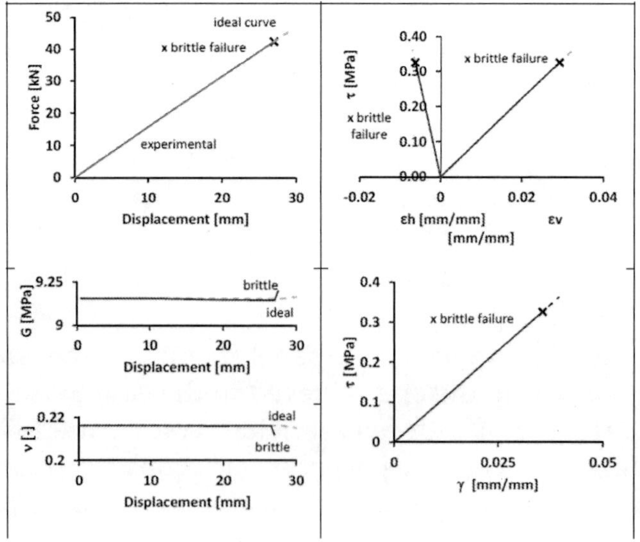

Figure 7 Cont.

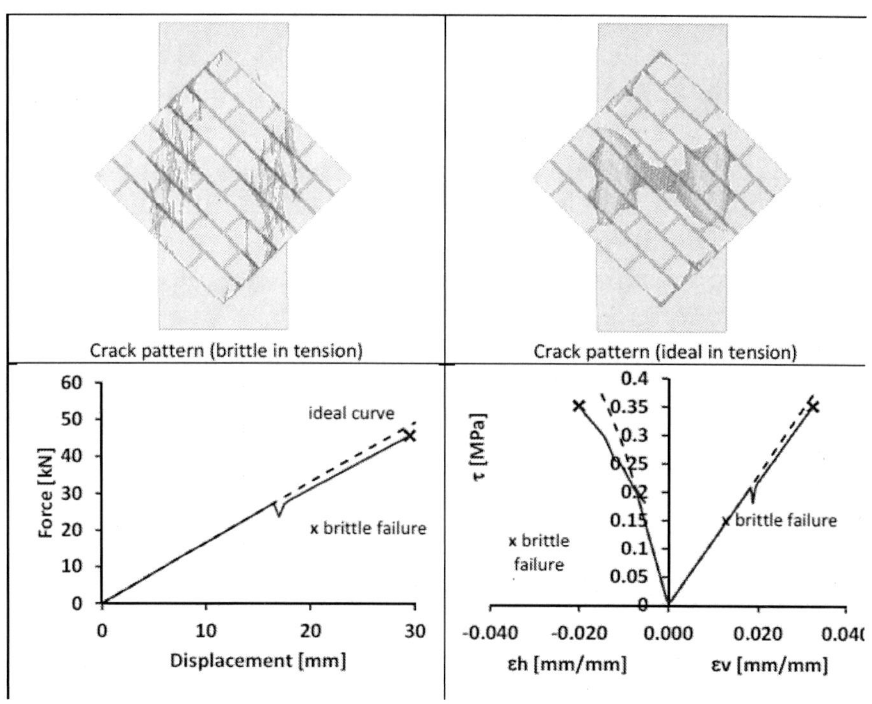

Figure 8 Numerical experimentation (reinforced LTC panel).

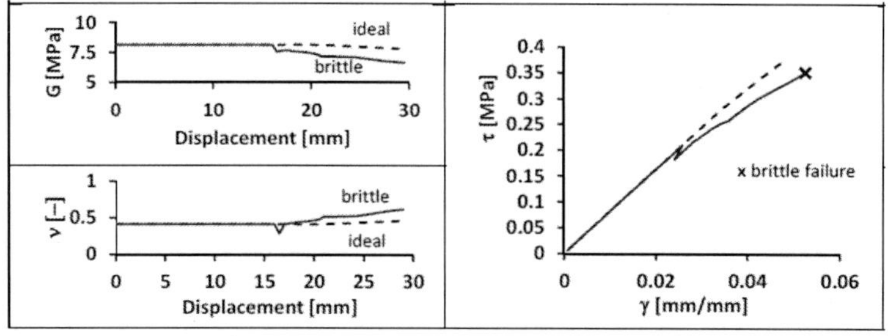

Figure 8 Cont.

FEM Models: Mortar B (MB)

The panels modeled with a better mortar, as well as the previous LTC panels were not tested in reality. However, this case has been analyzed to assess the influence of the strength of the mortar on the global behavior of the masonry panels. The MB panels are still almost homogeneous panels (having identical elastic moduli). However, the strength of the mortar is about 2.5 times higher, compared to plain soil mortar (0 day curing), while the brick properties are the same as the plain panels case. In the case of the unreinforced panel, both the crack pattern and the global response in terms of force/ displacement are almost similar compared to the unreinforced plain panel (Figure 9).

Nevertheless, the panel presents a higher ultimate load and shear strength. The comparison with the unreinforced plain panel test allows one to find, in general, a spreading of the crack pattern and higher global performances; however, an easier and more direct comparison can be made with the previous unreinforced long-term curing test. Actually, the main difference, compared to the previous test, is the strength of the brick (that in this case is lower than the previous case), and the global effect is a reduction of the shear strength (almost 20% smaller), being comparable to the reduction of the strength of the bricks.

In the reinforced MB panel, the global response in terms of force/ displacement and crack pattern is almost similar to the previous case of the long-term curing panel reinforced with mesh (Figure 10). The main reason for this result is the presence of the mesh reinforcement. Actually, the reinforcement makes less evident the benefits of the strength of the basic materials on the global behavior (and in particular, the properties of the mortar where the mesh reinforcement is placed). As well as in the previous LTC case, compared to the plain panel with mesh, the strength is doubled. Compared to the plain panel without mesh, the increase of strength is about 2.3 times, and the ultimate displacement is doubled.

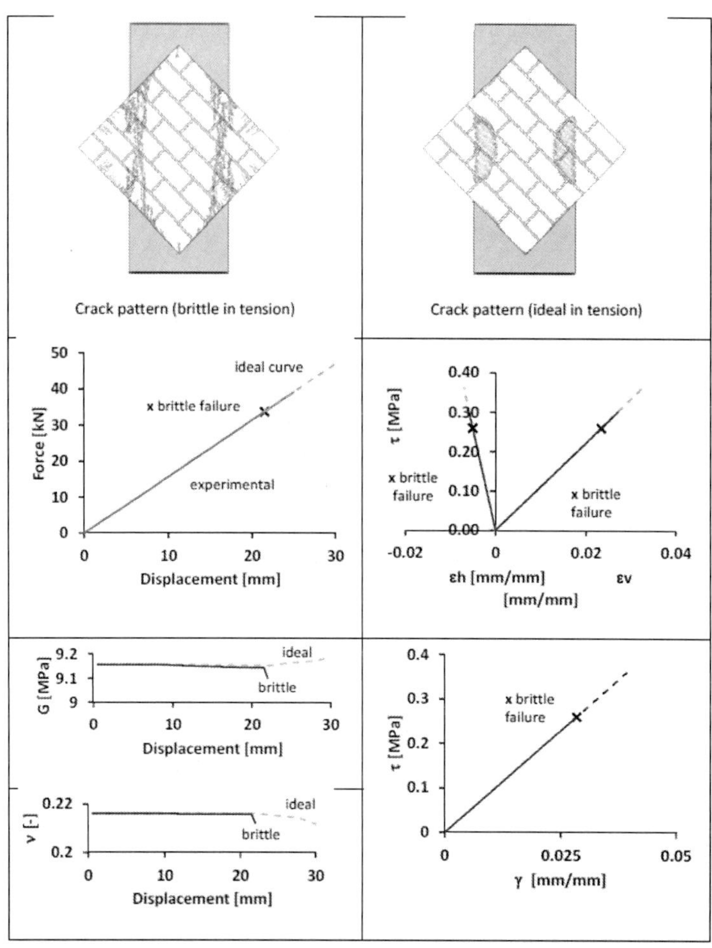

Figure 9 Numerical experimentation (unreinforced MB panel).

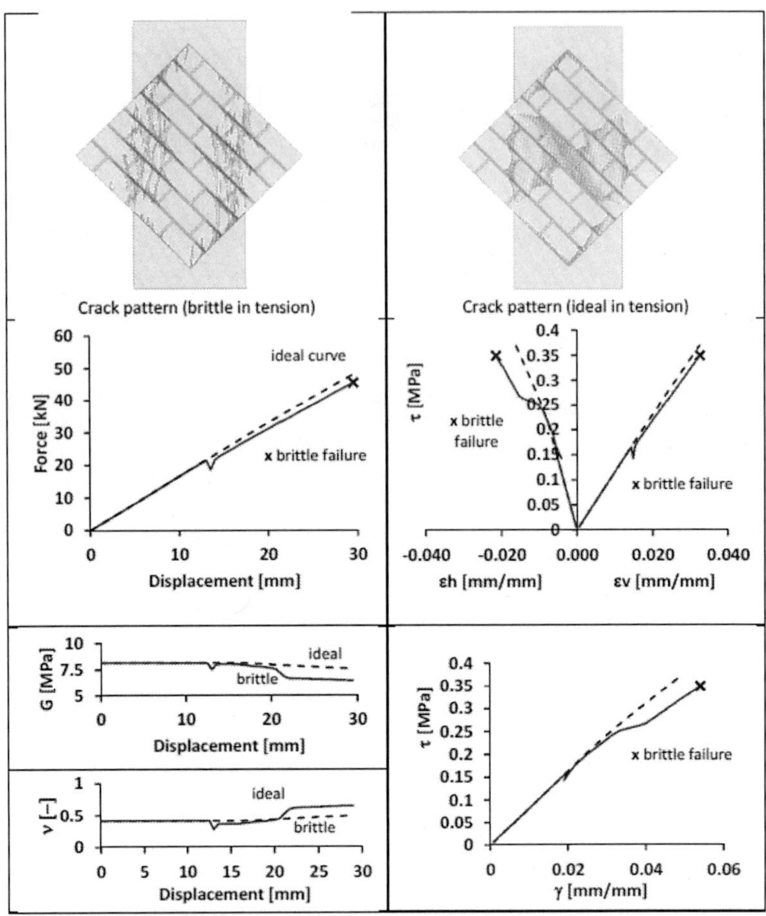

Figure 10 Numerical experimentation (reinforced MB panel).

CONCLUSIONS

Adobe earth constructions were copious in the ancient world. Furthermore, earth is still diffuse as a construction material, especially for its cheapness. Many historic structures, built for thousands of years, are now in need of conservation. In spite of their diffusion, only a few experimental tests are available. Actually, the variability of soil mechanic properties due to aging and composition strongly influence the seismic performance of adobe constructions. Plaster fiberglass mesh reinforcement represents a valid seismic reinforcement system for adobe building. The basic concept of this reinforcement is to improve the frictional resistance at the horizontal mortar joint location. In fact, the link between the brick units is the weakest section in the structural behavior of adobe walls. Numerical experimentation is a feasible way to deepen the knowledge of the seismic behavior of such structures.

After validating the numerical model, FEM simulations can be used as a tool to increase the knowledge of the effect of fiberglass mesh reinforcement, when varying the constituent materials, on global structural performances. The main scope of the present study is to highlight the influence of different mortar and brick compositions and aging combined with fiberglass mesh reinforcement on the in-plane shear performance of adobe walls. A literature survey clarified the effect of curing time on the strength of the soil material, namely an increase of strength both in tension and in compression. Then, numerical analyses allowed for us to remark on the following effects. The crack pattern is directly affected by an increase of the mortar strength; in fact, a stronger mortar is able to spread the smeared cracking strain field. In Table 3, ratios are evaluated in relation to the test on the unreinforced plain soil panel. The increase of the mortar strength (e.g., equal for both LTC and MB simulations), yet having the same stiffness of basic materials, leads to an increase of global shear strength. However the increase, at a global level (about 2.5 times), is nearly proportional to the mortar strength increase at the local level (about 2.5 times), but it is mitigated by the reduction of brick strength; in fact, a reduction of the brick strength of about 20% corresponds almost proportionally to a 20% reduction of the global shear strength. However, the presence of the fiberglass mesh reinforcement makes less evident the influence of the better mechanical properties of the mortar or bricks. The mesh is able to increase the shear strength of the panel, not altering its global stiffness. An interesting future development of this research (which is based on the reinforcement of adobe during construction) could be the extension of the study to the strengthening of existing structures by means of joint external repointing. In fact, the joint's reinforcement was proven to be effectively improving the adobe shear behavior.

Table 3 The main numerical outcomes (the ratios are related to the unreinforced plain panel).

FEM model	Material level (input data)		Global level (outcomes)		
	Brick strength ratio	Mortar strength ratio	G (MPa)	τ (MPa)	τ ratio
Plain (unreinforced)	1	1	9.17	0.15	1.00
Plain (reinforced)	1	1	8.14	0.19	1.26
LTC (unreinforced)	1.25	2.5	9.17	0.32	2.20
LTC (reinforced)	1.25	2.5	8.14	0.35	2.31
MB (unreinforced)	1	2.5	9.17	0.25	1.70
MB (reinforced)	1	2.5	8.14	0.34	2.31

ACKNOWLEDGMENTS

The analyses were developed within the activities of Rete dei Laboratori Universitari di Ingegneria Sismica—ReLUIS for the research program funded by the Dipartimento di Protezione Civile—Progetto Esecutivo 2010–2013.

REFERENCES

1. Revuelta-Acosta, J.D.; Garcia-Diaz, A.; Soto-Zarazua, G.M.; Rico-Garcia, E. Adobe as a sustainable material: A thermal performance. J. Appl. Sci. 2010, 10, 2211–2216.
2. Binici, H.; Aksogan, O.; Bakbak, D.; Kaplan, H.; Isik, B. Sound insulation of fibre reinforced mud brick walls. *Constr. Build. Mater.* 2009, *23*, 1035–1041.
3. Houben, H.; Guillaud, H. *Earth Construction a Comprehensive Guide*; Intermediate Technology Publications: London, UK, 1994.
4. Tolles, L.E.; Krawinkler, H. *Seismic Studies on Small-Scale Models on Adobe Houses*. The John A. Blume Earthquake Engineering Center, Department of Civil Engineering, Stanford University: Stanford, CA, USA, 1990.
5. *Standard Test Method for Diagonal Tension (Shear) in Masonry Assemblages*, ASTM E519-02; American Society for Testing Materials (ASTM): West Conshohocken, PA, USA, 1981.
6. Lignola, G.P.; Prota, A.; Manfredi, G. Nonlinear analyses of tuff masonry walls strengthened with cementitious matrix-grid composites. *J. Compos. Constr.* 2009, *13*, 243–251.
7. Turanli, L.; Saritas, A. Strengthening the structural behavior of adobe walls through the use of plaster reinforcement mesh. *Constr. Build. Mater.* 2001, *25*, 1747–1752.
8. Turanli, L. Evaluation of Some Physical and Mechanical Properties of Plain and Stabilized Adobe Blocks. Master's Thesis, Middle East Technical University, Ankara, Turkey, 1985.
9. Manie, J.; Kikstra, W.P. *DIANA Finite Element Analysis: User's Manual release*

9.4.3; Available online: https://support.tnodiana.com/manuals/d943/Diana.html (accessed on 13 February 2014).

10. Lignola, G.P.; Prota, A.; Manfredi, G. Numerical investigation on the influence of FRP retrofit layout and geometry on the in-plane behavior of masonry walls. *J. Compos. Constr.* 2012, *16*, 712–723.
11. Parisi, F.; Lignola, G.P.; Augenti, N.; Prota, A.; Manfredi, G. Rocking response assessment of in-plane laterally-loaded masonry walls with openings. *Eng. Struct.* 2013, *56*, 1234–1248.
12. Parisi, F.; Lignola, G.P.; Augenti, N.; Prota, A.; Manfredi, G. Nonlinear behavior of a masonry sub-assemblage before and after strengthening with inorganic matrix-grid composites. *J. Compos. Constr.* 2011, *15*, 821–832.

Chapter 13

NANOFIBERS REINFORCED POLYMER COMPOSITE MICROSTRUCTURES

Alubaidy[1], K. Venkatakrishnan[2] and B. Tan[3]

[1]School of Mechanical and Electrical Engineering, Sheridan Institute of Technology & Advanced Learning, Brampton, Canada.

[2]Department of Mechanical Engineering, Ryerson University, Toronto, Ontario.

[3]Canada Department of Aerospace Engineering, Ryerson University, Toronto, Ontario, Canada.

INTRODUCTION

In general, Nano composites are defined as the combination of multiphase materials in which at least one of the constituents has one dimension in the nanometer range [1]- [3]. The Nano scale constituent could be one dimensional like nanowires and nanowires, two-dimensional like Nano clay or three-dimensional like spherical particles in Nano scale range. Nano fibers reinforced polymer multi functionality can be attributed to the combination of the constituent materials. Desired properties of Nano fibers reinforced polymer can be obtained by the selection of the constituent materials and the size of the nanowires based on the required application. Current research has focused in the areas of manufacturing techniques and material combination for the fabrication of the nanostructured reinforced polymers [4], [5].

Nano fibers reinforced polymer are progressing with the use of a combination of atomic scale characterization and detailed modeling. In the early 1990s, Toyota Central Research Laboratories in Japan reported working on a Nylon-6 Nano composite [6], in which a small amount of Nano filler resulted in a considerable improvement of thermal and mechanical properties. The properties of Nano fibers reinforced polymer materials depend on their morphology and interfacial characteristics as well as on the properties of their individual parents (Nano fillers and polymer, in this case).

Dramatic changes in physical properties will be the result of the transition from micro particles to nanoparticles. Nano scale materials have a large surface area for a given volume [7]. A nanostructured material can have substantially different properties from a larger-dimensional material of the same composition because many important chemical and physical interactions are governed by surfaces and surface properties. In the case of nanoparticles and Nano fibers, the surface area per unit volume is inversely proportional to the material's diameter. So, the smaller the diameter, the greater is the surface area per unit volume [7]. Figure 1 shows common particle geometries and their respective surface area-to-volume ratios. For the Nano fiber and layered material, the surface area to volume ratio is dominated by the first term in the equation, especially for Nano materials. The second term (2/l and 4/l) has a very small influence and is often omitted compared to the first term. Therefore, a change from the micrometer to nanometer in particle diameter, layer thickness, or fibrous material diameter range, will affect the surface area to volume ratio by three orders of magnitude [8].

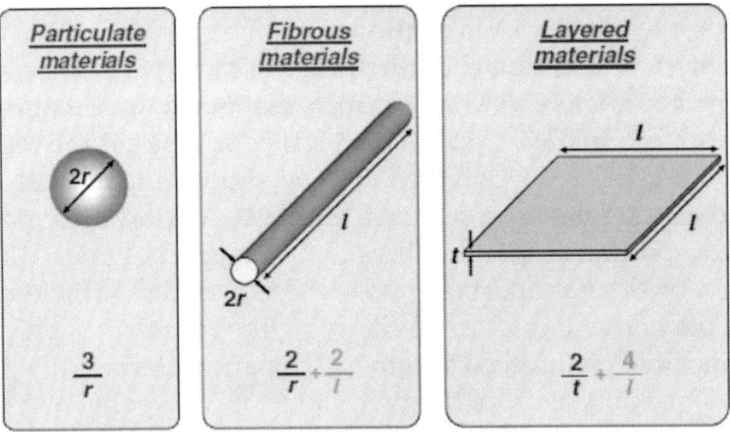

Figure 1. Common particle reinforcements/geometries and their respective surface area to volume ratios. [8].

Typical Nano materials currently under investigation include nanoparticles, nanotubes, Nano fibers, fullerenes, and nanowires. In general, these materials are broadly classified by their geometries [9]: particle, layered, and fibrous Nano materials [8], [9]. Carbon black, silica nanoparticle, polyhedral oligomericsislesquioxanes, can be classified as nanoparticle reinforcing agents, while Nano fibers and carbon nanotubes are examples of fibrous materials [9]. When the filler has a nanometer thickness and a high aspect ratio (30–1000) plate-like structure, it is classified as a layered nanomaterial such as an organ silicate [10].

In general, the high aspect ratio of Nano materials provides the necessary reinforcement properties. The properties of reinforced polymers are greatly influenced by the size of its nanomaterial and the quality of the interfacing between the matrix material and the filler material. Significant differences in composite properties may be obtained depending on the nature of the filler material used whether it's layered silicate or Nano fiber, action exchange capacity, or polymer matrix and the method of preparation [11]. As an example, when the polymer is unable to intercalate (or penetrate) between the silicate sheets, a phase-separated composite is obtained, and the properties stay in the same range as those for traditional micro composites [10]. In an intercalated structure, where a single extended polymer chain can penetrate between the silicate layers, a well-ordered multilayer morphology results with alternating polymeric and inorganic layers. An exfoliated or delaminated structure is obtained when the silicate layers are completely and uniformly dispersed in a continuous polymer matrix. In any case, the physical properties of the resultant Nano fibers reinforced polymer will be significantly different, as discussed in the following sections. Similarly, in fibrous or particle-reinforced polymer Nano fibers reinforced polymer (PNCs), dispersion of the nanoparticle and adhesion at the particle–matrix interface play crucial roles in determining the mechanical properties of the Nano fibers reinforced polymer. The nanomaterial will not offer improved mechanical properties without proper dispersion. A poorly dispersed nanomaterial may degrade the mechanical properties of the produced reinforced polymers [12]. Additionally, optimizing the interfacial bond between the nanostructures and the matrix, one can tailor the properties of the overall Nano fibers reinforced polymer in a similar manner to what is done in macro composites. As an example, good adhesion at the interface will improve properties such as inter laminar shear strength, fatigue, delamination resistance, and corrosion resistance. Finally, it is important to recognize that Nano fibers reinforced polymer researches are extremely broad, encompassing areas such as communications, electronics and computing, data storage, aerospace and sporting materials, health and medicine, transportation, energy, environmental, and many other applications. The focus of this chapter is to highlight the state of knowledge in processing, fabrication, characterization, properties, and potential applications of the Nano fibrous reinforced polymer microstructures

FABRICATION TECHNIQUES

The advancement in the technology of Micro Electro Mechanical Systems (MEMS) has demand for the fabrication of 3D micro/nanostructures and devices. The excitement surrounding the Nano scale science and technology gives us unique opportunities to develop and examine revolutionary processes and materials. Nano fibers reinforced polymer embedded with 2D and 3D micro/Nano materials find many applications in the field of medicine, tissue engineering, drug delivery, antibacterial implants or catheters, modification of textiles, and modification of polymers. Many optical, electrical and magnetic applications, have opened up new areas of research for manufacturing Nano fibers reinforced polymer with engineered nanoparticles materials.

Researchers have been working on different micro/nano manufacturing techniques. LIGA (German acronym for Lithographie, Galvanoformung, Abformung) [7], [8], Photolithography (6), Electrochemical Fabrication (EFAB) [9], localize electrochemical deposition [10] and laser sintering [13] are some of the techniques used for micro/nano fabrication. Some of these techniques have been used for the fabrication of nanofibers reinforced polymer by dispersing Nano fibers in polymer resins.

The LIGA technique was invented approximately 20 years ago. It is a powerful method that facilitates the high volume production of Nano fibers reinforced polymer components for many fields of applications [14]. Researchers presented designs, fabrication and experimental results of high power electrostatic micro actuators, using LIGA process. This process is capable of producing high aspect ratio microstructures of Nano fibers reinforced polymer, as shown in Figure 3 [15].

The LIGA process uses the simple shadow printing process onto a resist on an electrically conducting substrate. After development of the irradiated resist, an electroforming step fills the holes of the relief with metal. This more stable body is used as a mold insert for further molding or embossing steps [8]. The deep lithography step, performed by the use of highly parallel and collimated synchrotron radiation, is the basic step, thus not only defining the shape but also the structural accuracy of the final product. [16].

Typically LIGA structures allow for the free choice of the lateral 2D pattern that is projected into the third dimension to form prismatic or cylindrical geometries. Generally, this technique has been used to produce structures with straight walls. However, for all major LIGA process steps, variations have been developed to increase the fabrication flexibility. Geometrical variations in the third dimension (vertical) are possible and can be obtained in different ways by modifying or combining process steps, in particular for producing shapes with increased dimensionality. The consecutive process steps of deep UV lithography, electroforming, and plastic molding can be used to fabricate three dimensional (3D) microstructures with almost no restrictions in their lateral shape in a large variety of Nano fibers reinforced polymer.

Photolithography is one of the widely used techniques for fabrication of Nano fibers reinforced polymers. Two photolithography-based approaches has been presented to directly micro machine photo pattern able upper hydrophobic micro patterns with excellent adaptability and flexibility to a wide variety of substrates, employing the Nano morphology and hydrophobicity of polytetrafluoroethylene (PTFE) nanoparticles and the photo pattern ability and transparency of an SU-8 photoresist [17].

A light source such as UV light is used in photolithography to polymerize the photo-responsive resin with suspended nanoparticles through a mask [6]. Depending upon the requirement of the feature, a mask is prepared and the feature is transferred using a UV source light. The removal of the unaffected part by the light source is performed using the secondary process of chemical etching. UV light has been used primarily for transferring the mask, but various other alternatives such as X-ray lithography, lithography, Nano imprint and ion projection lithography are being explored [6], [18]. For several decades, this technique has been researched and developed, but the resolution of the fabricated feature using photolithography is limited due to the optical diffraction limitation. Photolithography also requires the use of expensive masks and molds for the fabrication process. These techniques are effective for the mass fabrication of high resolution micro features, but they are limited to 2D geometries [18]. Nano fibers reinforced polymer microstructures can be fabricated by dispersing the photosensitive resin with nanomaterial's [14]. Figure 2 shows SEM photographs of the photolithographic patterns for photosensitive polyimide with montmorillonitenanofibers reinforced polymers [19].

Electrochemical Fabrication (EFAB) was originally invited to addresses the long development time for Optical MEMS, which can go up to few weeks. This method has also been used for the fabrication of Nano fibers reinforced polymer micro structures. The fabrication of Nano fibers reinforced polymer film of polypyrrole (PPY) and TiO_2 nanotube (TNT) arrays via electrochemical methods was reported in several articles. A novel dual-layered photoreceptor based on the reinforced polymer film as a charge generation layer (CGL) was designed and fabricated [20].

EFAB is a solid free-form fabrication technology that creates complex, miniature three-dimensional shapes based on 3-D computer aided design CAD data [8]. Inspired by rapid prototyping methods, EFAB can fabricate complex shapes by stacking multiple patterned layers. Unlike rapid prototyping, EFAB is a batch process that is suitable for volume production of fully functional devices in engineering materials, not only models and prototypes [9]. EFAB has some limitations and shortcomings. EFAB is a technique that requires the use of masks to build two-dimensional (2D) planar structures. High aspect ratio microstructures are thus a result of multiple steps of material deposition or removal through the use of several masks, that requires increased fabrication time and cost.

A conducting microelectrode is utilized in a localized electrochemical deposition (LECD) technique to fabricate high aspect ratio structures. During fabrication, a localized deposition is produced by placing an electrode tip that has micro meter-scale dimensions, near a substrate in an electrolyte and applying an electric potential between them. Confined deposition is produced due to the highly localized electric field in the region between the microelectrode and the substrate. High aspect ratio microstructures result from the displacement of the end of the electrode along the trajectory of the desired geometry while maintaining continuity with the deposited materials. However, Nano composite micro structures fabricated by this method are usually porous and have feature sizes in the tens of micrometers due to the limitation in fabricating and maintaining a sharp conductive probe, and in confining the electric field down to the Nano scale dimensions [10].

Laser sintering is another technique that has been widely used for the micro fabrication of Nano fibers reinforced polymers. In this method, a high intensity laser is used to ablate the material and form nanoparticles. Chen et al, reported a new technique for conductive Nano fibers reinforced polymer micro fabrication and they were able to lower the percolation threshold of the Nano fibers reinforced polymer [21]. This process takes place inside a liquid polymer resin which is then polymerized using UV light to form Nano fibers reinforced polymers layers. The main advantage of this technique is that metals, non-metals, glass, polymers can be utilized to fabricate Nano fibers reinforced polymer s with various properties. The main drawback of this technique is its limitation to planar geometries. Stacking of different layers has been tried, but the alignment and handling of the layers contributes to the limitation of higher aspect ratio of the micro device [13].

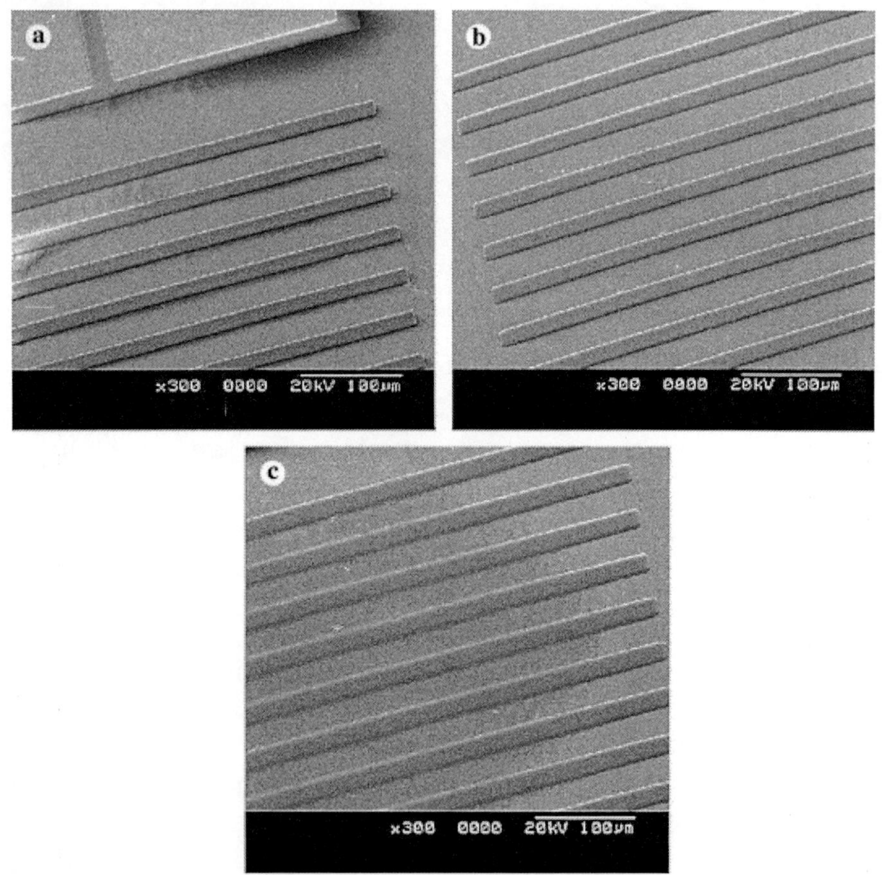

Figure 2. SEM photographs of the photolithographic patterns of (a) pure PSPI, PSPI/MMT Nano fibers reinforced polymers with (b) 2 wt. % and (c) 3 wt. % MMT contents. [19].

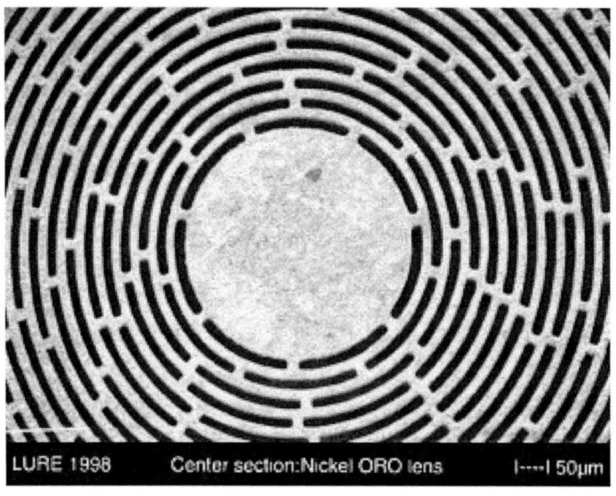

Figure 3. Center section of a fabricated ORO lens. Nickel matrix with 20 μm wide and 600 μm deep holes. Measured aspect ratios (sidewall inclination) >400 [15].

The nonlinear optical process of multi photon absorption (MPA) was first predicted in 1931 by the Nobel laureate physicist Marie Goeppert-Mayer in her doctoral dissertation [22]. The technique was not verified experimentally until the advent of laser. An intuitive explanation of MPA is the transition from the ground electronic state to an excited electronic state which is usually achieved by the absorption of a high-energy photon or instead reached by simultaneous absorption of multiple low energy photons. The most common implementation of this method is degenerate two-photon absorption (TPA) where both photons have the same energy. The spatial confinement of MPA excitation is used to induce a chemical reaction at the laser focal point. This polymerization reaction occurs by radical reaction mechanisms that depend on the photo initiator and monomer being used. The photo initiators used for radical MPA vary from small molecules to large conjugated molecules. A number of groups have reported the successful application of radical MPA using a different kind of resins, homemade and commercial, and different excitation sources. Custom photo initiators have been designed by several groups and have been shown to be effective both for low threshold powers and for the ability to use less expensive laser systems. While the benefits of custom initiators are clear, their availability is limited. Commercial resins or resins made of commercial components have the advantage of accessibility but suffer from a slightly higher power threshold for fabrication. However, for the entire laser systems used, the threshold for these resins is always well below the available power and therefore their use is completely practical.

MPA can achieve resolution that is considerably better than that predicted by the diffraction limit due to a combination of optical nonlinearity. The probability for MPA is proportional to I_n, where I is the light intensity and n is the number of absorbed photons. This effectively narrows the point-spread function (PSF) of the beam near the focal point so that it is smaller than the diffraction limit at the excitation wavelength. The real benefit of the optical nonlinearity of MPA lies in the negligible absorption away from the focal point. Photo initiator concentrations can be employed that are about ten times more than would be feasible for single-photon excitation without any fear of out-of-plane polymerization.

Multi Photon Absorption (MPA) is a relatively new and evolving method for the manufacturing of micro/Nano structures. Unlike other methods, this technique does not involve any secondary operations or processes for Nano fibers reinforced polymer fabrication. Micro/Nano structures and devices are fabricated with higher accuracy and complexity using MPA than other prevailing methods of fabrication for various applications. When an ultra-short laser pulse is focused in a photo responsive polymer resin, a solid voxel (volumetric pixel) is generated. The voxel size defines the minimum resolution of the polymer which is converted into solid form. Complex 2D and 3D microstructures can be fabricated by scanning the laser in the photo responsive resin [23]. Research on fabrication of micro/Nano structures from different types of polymers using MPA has been conducted in the past decades. In certain applications there is a real need of complex Nano fibers reinforced polymer structures fabrication [24]. The prevailing methods for Nano fibers reinforced polymer manufacturing are good for planar or 2D device fabrication. For mask less manufacturing 2D and 3D Nano fibers reinforced polymer devices with accuracy and precision there is a need of identifying or performance research towards the development of new mask less fabrication techniques using MPP due to its advantages over other methods.

NANOFIBERS REINFORCED POLYMERS ENHANCED PROPERTIES

The samples that presented in this section are polymer microstructures fabricated with MPA using laser ablative synthesis generated Nano fibers as reinforcement [25] [26]

Mechanical Properties

Nano Indentation Tester (Figure 4) uses an already established method where an indenter tip with a known geometry is driven into a specific site of the material to be tested, by applying an increasing normal load. When reaching a pre-set maximum value, the normal load is reduced until partial or complete relaxation occurs. At each stage of the experiment, the position of the indenter relative to the sample surface is precisely monitored with a differential capacitive sensor. For each loading/unloading cycle, the applied load value is plotted with respect to the corresponding position of the indenter. The resulting load/displacement curves provide data specific to the mechanical nature of the material under examination. Established models are used to calculate quantitative hardness and elastic modulus values for such data [27]. Hardness and elastic modulus can be determined using the method developed by Oliver and Pharr [28]. The hardness represents the resistance of a material to local surface deformation. The Indentation Testing Hardness, H, is determined from the maximum load, F_{max}, divided by the projected contact area A_p at the contact depth h_c [29];

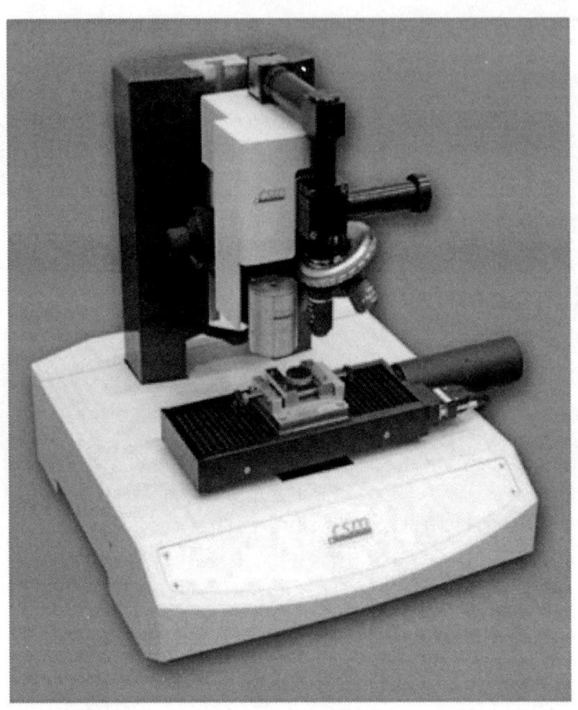

Figure 4. Nano Indentation tester (CSM Instruments)H=FmaxAp (hc)

Where A_p (h_c) is the projected area of indenter contact at distance, h_c from the tip and h_c is the depth of the contact of the indenter with the test piece at F_{max}. Which can be expressed by *hc=hmax− (hmax−hr)* , where h_{max} is the maximum indentation depth at F_{max}, h_r is the point of intersection of the tangent to the unloading curve with the depth axis, and ε is a constant depending on the nonlinearity of the unloading curve (Figure 5). For modified Berkovich indenter, *Ap* (*hc*) =24. 5 *h2c* [30]. The elastic modulus represents the overall stiffness of the reinforced polymer network and the reduced modulus of the indentation contact, E_r, is given by [29];

*Er=*ϖ√. *S2Ap* (*hc*) √

where S is the contact stiffness or the slope of the unloading curve shown in Figure 5 at the point of maximum load which can be found by [31];

S=Fmaxhmax−hr

The relationship between Indentation Modulus, E, and the

reduced modulus, E_r, of the sample is given by [29];

$$1Er=1-\ 2sE+1-\ 2iEi$$

Where v_i is the Poisson's ratio of the indenter, v_s is the Poisson's ratio of the sample and E_i is the modulus of the indenter. The elastic stress–strain characteristics of a randomly Nano fibers dispersed polymer are expressed by three elastic constants, namely, Young's modulus E, Poisson's ratio n vs, and shear modulus G. Only two of these three elastic constants are independent since they can be related by the following equation [32];

$$G=E_2\ (1+vs)$$

By substituting the values of E, and $v_{s,}$ the shear modulus can be calculated. A thin layer containing randomly oriented discontinuous nanofibers exhibits planar isotropic behavior. The properties are ideally the same in all directions in the plane of the layer. For such a layer, the tensile modulus and shear modulus are calculated from [32];

$$E=38E11+58E22\ G=18E11+14E22$$

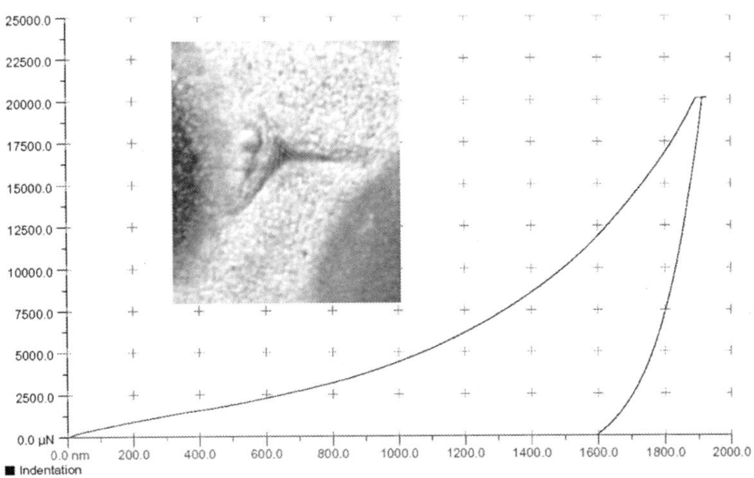

Figure 5. Typical indentation load-displacement curves for the reinforced polymer

where E_{11} and E_{22} are the longitudinal and transverse tensile moduli for a unidirectional discontinuous Nano fiber lamina of the same Nano fiber aspect ratio, and same Nano fiber volume fraction as the randomly oriented discontinuous Nano fiber composite. Rearrange Eqn 7 and subtract Eqn 6 from Eqn 7, then using transverse modulus $E22=1+2\,tvf1-tvfEm,\ \ t=(Ef/Em)-1\ (Ef/Em)+2$, Poisson's ration $s=vfvf+vmvm$ and $vf+vm=1$, the volume fraction ratio of the Nano fibers in the polymer matrix can be obtained by [31];

$$vf=8(3G-E)Em-1\ (Ef/Em)-1(Ef/Em)+2\ 2+8\ (3G-E)\,Em$$

Where v_m is the volume fraction of the polymer matrix, v_f and v_m are Poisson's rations of Nano fibers and polymer matrix respectively, E_f and E_m are Young's moduli for Nano fibers and polymer matrix respectively. By substituting the appropriate values in Eqn 8, the volume fraction of the Nano fibers in the polymer matrix can be calculated. An important function of the matrix in a Nano fiber-reinforced composite material is to provide lateral support and stability for Nano fibers under longitudinal compressive loading like the one applied by the indenter head. In polymer matrix composites with which the matrix modulus is relatively low compared to the Nano fiber modulus, failure in longitudinal compression is usually initiated by localized buckling of Nano fibers. In general, the shear mode of failure is more important than the extensional mode of failure and the shear mode is usually controlled by the matrix shear modulus as well as Nano fiber volume fraction.

A viscoelastic polymer has both elastic component and a viscous component. Applying stress to a polymer causes molecular rearrangement due to changing positions by parts of the long polymer chain (creep). Polymers remain solid even when these parts of their chains are rearranging in order to accompany the stress which creates a back stress in the material. The material no longer creeps when the back stress is of the same magnitude as the applied stress. If the original stress is taken away, the accumulated back stresses will cause the polymer to get back to its original form with the help of its elastic component. The polymer loses energy when a load is applied then removed with the area between loading and unloading curves (Figure 5) being equal to the energy lost during the loading cycle. Thus permanent plastic deformation of the polymer occurs even when the applied load is removed.

The large surface area of Nano fibers provides better interaction between the polymer chains and Nano fibers. The Nano fibrous structures act as a preferential nucleation site for crystalline phases which cause an increase in the modulus and hardness. Hardness, which is directly related to the flow strength of a material, depends on the effective load transfer between the matrix and the reinforcement phase in the Nano fibers reinforced polymer. Strong interfacial bonding between the matrix and the Nano fibers is essential for efficient load transfer to obtain high strength. Plastic deformation in polymers occurs by nucleation and propagation of shear bands which in the unreinforced polymer matrix propagate unhindered as there are no barriers for their movement. Thus, the presence of Nano fibers in the Nano fibers reinforced polymers could offer resistance for the propagation of shear bands. A good mechanical interlocking and the presence of obstacles to the motion of shear bands are the reasons for the enhancement of hardness and elastic modulus in polymer dispersed Nano fibers.

ELECTRICAL CONDUCTIVITY

Many techniques for the synthesis of conductive polymers have been developed. Most conductive polymers are prepared by oxidative coupling of monocyclic precursors. One challenge is usually the low solubility of the polymer. An important property of the electrically conductive reinforced polymer is the positive temperature coefficient (PTC) effect. PTC reflects the increase in electrical resistivity of the composites during the heating process, and hence decreasing the electrical conductivity of the reinforced polymer. PTC materials have so many potential applications, including sensors, self-regulating heaters, and switching materials. The PTC behavior occurs as a result of the difference in the coefficients of thermal expansion between the fillers and the matrix.

Electrical properties of compositions made of Nano fibers, dispersed in polymers, are mainly characterized by the formation of a conduction network by contact conditions between neighboring Nano fibers in the network. Carbon Nano fibers (CFN's) dispersed in the developed resin was used as a conductive composition materials in several applications. The conductivity of the Nano fibers reinforced polymer films can be measured using two points probe testing device from SVSL abs Inc as shown in Figure 6. Four probe systems can also be used. The voltage passing through the Nano fibers reinforced polymer film can be recorded, as well as the current and the value of the resistance (and hence the conductivity) can be obtained using Ohm's law.

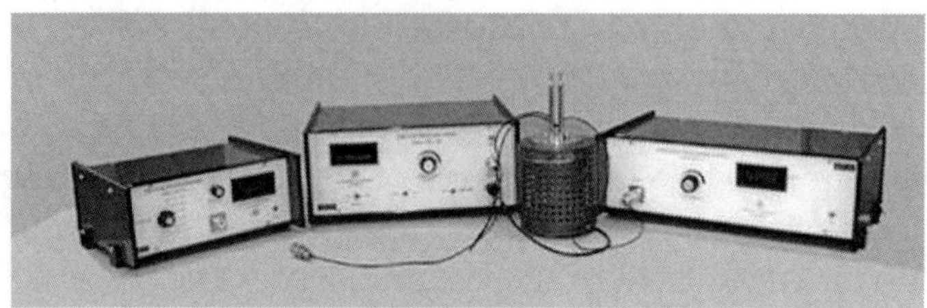

Figure 6. Two points probe testing device (SVSLabsInc).

Electrical conductivity in CNF's reinforced polymer is generated as a result of the direct contact between CNF's and tunneling resistance determined by the width of the insulating resin around the CNF's. Thermal expansion caused by heat gradient increases gap width between contiguous CNF's and reduces the number of conductive pathways which results in a decrease in electrical conductivity. The temperature dependence of the electrical conductivity can be explained by the general theory of the thermal fluctuations. The resistivity of the junction is given by;

$$= oe - T1/\ (T+To)$$

Where the constants ρ_0, T_1, and T_0 depend essentially on the characteristics of the tunnel junctions, which are supposed to be functions of various parameters such as filling factor, filler size and shape, sample processing. The relative resistivity (ρ_r) can be used to characterize the intensity of the PTC effect:

$$r=\log(\ 140\ 20)$$

Where ρ_{140} and ρ_{20} are the resistances of the composites at two different temperatures, for example, 140 and 20°C. Thus the relative conductivity can be expressed as;

$$Sr=\log(\ 20\ 140)$$

This represents how sensitive and to what extent the reinforced polymer resistance responds after being stimulated by the temperature change. The electrical sensitivity of a reinforced polymer fabricated by femtosecond laser material processing can be expressed by;

$$Ks=\Delta Sr\Delta\ r$$

Where K_s is the electrical sensitivity, ΔS_r is the change in relative conductivity and $\Delta\omega_r$ is the change in relative repetition rate of the femtosecond laser that can be expressed by;

$$r=\ min$$

Where ω_{min} is the minimum repetition rate used for the generation of Nano fibers. For example, The equation of the straight line obtained from Figure 7 is Sr=0. 17 r−10. 61 17 which indicate that the electrical sensitivity Ks is about 0. 17. The higher the value of the electrical sensitivity, the more sensitive is the reinforced polymer to change in temperature and thus less conductivity at higher temperatures. Also, the PTC effect of Nano fibers reinforced polymers fabricated with higher laser repetition rate is weaker than that of Nano fibers reinforced polymer s fabricated with lower repetition rate. This is a direct result of the matrix volumetric expansion which cause more contiguous CFN's to get disconnected and reduce number of electrical pathways.

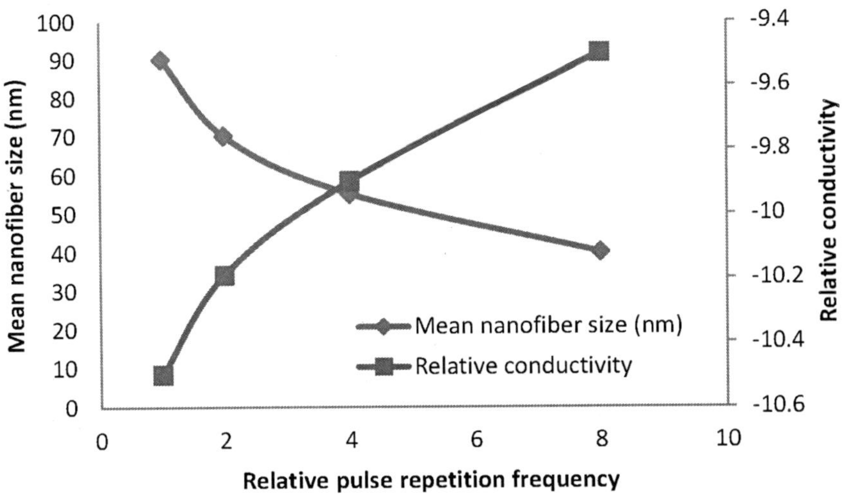

Figure 7. Pulse repetition frequency effect on the relative conductivity of the Nano fibers reinforced polymer microstructures.

MAGNETIC ENHANCEMENT

Magnetic neodymium-iron-boron (NdFeB) Nano fibers and nanoparticles, have become one of the most important spot in the research field of magnetic nanomaterial's to meet the demand for miniaturization of electronic components recently. They have been successfully prepared by various techniques like the sol-gel auto-combustion method [33], co-precipitation [34], hydrothermal method [35], reverse micelles [36], micro emulsion method [37], alternate sputtering [38], pulsed laser deposition [39], and many others.

The total magnetization of a Nano fiber is given by the vectorial sum of all single magnetic moments of the atoms. As for the atomic magnetic moments in Nano fibers, the average magnetization will be zero in the absence of magnetic field since all magnetic moments are randomly directed in space. When a magnetic field is applied by the substrate, the magnetic moments orient in the direction of the field and give rise to a net magnetization of the Nano fibers [40].

In ferromagnetic materials (Figure 8 (a)), the saturation magnetization MS of the magnetic Nano fibers reinforced polymer microstructures at room temperature is higher than that of the pure polymer. The larger cervicitis HC that we see could be as a result of the oxide Fe. Figure 8 (b) shows the room temperature M-H curves for the polymer (Ormocer). It has a weak diamagnetic response to the applied field, but appears to have a small amount of unknown impurity which contributes to a very weak non- linear deviation near zero fields that could be a paramagnetic effect. Small traces of paramagnetic impurities can be found in almost all materials and hence, it is not surprising in these polymers. However, these trace impurities do not affect the quality of our ferromagnetic composites. Figure 8 (c)shows the magnetic measurements of the Nano fibers reinforced polymers which has a coercive field HC of approximately 260 Oe at room temperature. This large coercively suggests that oxide Nano fibers are present on the microstructure.

It can be seen that the magnetization of the generated Nano fibers reinforced polymer increases compared to pure polymer. This increase in magnetization is expected below the super paramagnetic-ferromagnetic transition temperature, which is above 300K for the magnetic nan fibers of 20 nm average size due to reduced thermal activation energy. Nanoparticle interactions, which depend on the iron concentration in the polymer matrix, strongly influence the remnant magnetization. Since agglomeration of nanoparticles into nan fibers is observed in all our polymer Nano fibers reinforced polymer samples, interactions are expected to play a significant role in the magnetic response. These interactions lead to a non- linear increase in MR as the concentration of iron is increased. The magnetic interactions are generally expected to be dipolar in nature, although in strongly coupled clusters, exchange interactions are also possible.

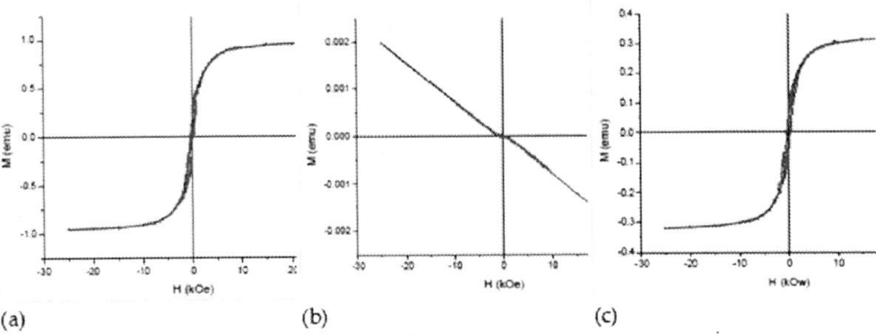

Figure 8. Room temperature M-H curves (a) Magnetic Nan fibrous structures, (b) Ormocer, and (c) magnetic Nano fibers reinforced polymer.

APPLICATIONS

Bioapplications

There are unlimited applications of reinforced polymer microstructures. One of the most recent application is DNA sensing based on reinforced polymer micro structuring. A conducting polymer containing Nano fibrous structure can be very sensitive even to small perturbations at the interface, thanks to its high electrical conductivity and electron transfer capabilities when the interface is covered by its film. Biocompatible conducting polymers can be variably modified for immobilization of probe DNA via covalent linking or electrostatic interactions. The conductivity of the conducting polymer electrochemically deposited on the electrode surface can be modulated by changing the pH of the medium, the electrochemical potential, and/or the electrolyte. Because of these characteristics and other advantages, conducting polymers could be utilized extensively for the construction of biosensors including DNA sensors which, to the best of our knowledge, is reported for the first time.

Noble metals are known to create strong chemical bonds with compounds terminating with a thiol (-SH) group. In order to achieve immobilization of nucleic acids on solid substrates, researchers have developed techniques by which DNA molecules can be linked to a thiol group. In this case, the throated terminal of molecules is chemisorbed on the substrate, while the DNA portion of the molecules are standing parallel to each other and away from the substrate. Since they are formed quickly, resulting in a well-defined and reproducible surface, and are stable under normal laboratory conditions, gold electrodes are fabricated using femtosecond laser material processing or other techniques [41]

Energy Applications

Reinfoced polymer micro structuring can be utilized in numerous application. Researchers prepared dye-sensitized solar cells using micro/Nano fibers reinforced polymer $Ti0_2$ porous films [42]. This result in cells with enhanced light collection. They applied a technique which opens an alternative way for manufacturing solar cells on an industrial scale. $Ti0_2$ micro/Nano-composite structured electrodes for quasi-solid-state dye-sensitized solar cells [43]. These revolutionary Nano-structured ultra-thin film solar PV products will provide affordable clean renewable energy for everyone. Another unique technology has been developed that absorbs and converts more sunlight throughout the day by utilizing special kind of Nano fibers reinforced polymers. This result in a dramatic increase in total power output. Each Nano fiber increases the total PV surface area by an incredible 6-12 times over current other thin film products on the market today. Figure 9 shows luminescent solar concentrators (LSCs) comprising CdSe core/multi shell quantum dots (QDs) developed by Bomm et al [44].

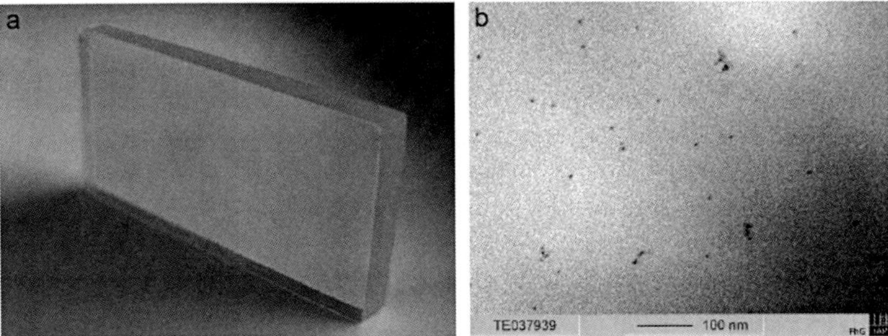

Figure 9. (a) Photograph of a P (LMA-co-EGDM) plate containing CdSe core/ multi shell QDs (illuminated by a UV-lamp) illustrating the concentrator effect and (b) TEM image of a QD-LSC/P (LMA-co-EGDM) Nano fibers reinforced polymer showing single QDs and a few small QD aggregates [44]

SUMMARY

In conclusion, this chapter focused on Nano fibers reinforced polymer microstructures, including fundamental properties, manufacturing techniques, and applications. The chapter also discussed the scientific principles and mechanisms in relation to the methods of processing, manufacturing and commercial applications. The mechanical, electrical, and magnetic properties of Nano fibers reinforced polymers has been discussed in details and it offers insight studies on technology, modeling, characterization, processing, manufacturing, and applications for Nano fibers reinforced polymer Nano composites.

REFERENCE

1. T. Rogers-Hayden and N. Pidgeon, "Moving engagement 'upstream'? Nanotechnologies and the Royal Society and Royal Academy of Engineering's inquiry," Public Understanding of Science, vol. 16, no. 3, pp. 345-364, Jul. 2007.
2. A. Pomogailo, "Synthesis and intercalation chemistry of hybrid organo-inorganic nanocomposites," Polymer Science Series C, vol. 48, no. 1, pp. 85-111, 2006.
3. J. -J. Luo and I. M. Daniel, "Characterization and modeling of mechanical behavior of polymer/clay nanocomposites," Composites Science and

Technology, vol. 63, no. 11, pp. 1607–1616, Aug. 2003.

4. E. Thostenson, C. Li, and T. Chou, "Nanocomposites in context," Composites Science and Technology, vol. 65, no. 3–4, pp. 491–516, Mar. 2005.
5. M. Alexandre and P. Dubois, "Polymer-layered silicate nanocomposites: preparation, properties and uses of a new class of materials," Materials Science and Engineering: R: Reports, vol. 28, no. 1–2, pp. 1–63, Jun. 2000.
6. J. Li and L. S. Melvin, "Sub-resolution Assist Feature Modeling for Modern Photolithography Process Simulation," Japanese Journal of Applied Physics, vol. 47, p. 4862, 2008.
7. A. Bertsch, H. Lorenz, and P. Renaud, "Combining microstereolithography and thick resist UV lithography for 3D microfabrication," in Micro Electro Mechanical Systems, 1998. MEMS 98. Proceedings., The Eleventh Annual International Workshop on, 1998, pp. 18–23.
8. A. Schmidt and W. Ehrfeld, "Recent developments in deep x-ray lithography," Journal of Vacuum Science&Technology B: Microelectronics and Nanometer Structures, vol. 16, no. 6, pp. 3526–3534, Jan. 1998.
9. A. Cohen, G. Zhang, F. G. Tseng, U. Frodis, F. Mansfeld, and P. Will, "EFAB: rapid, low-cost desktop micromachining of high aspect ratio true 3-D MEMS," in Micro Electro Mechanical Systems, 1999. MEMS'99. Twelfth IEEE International Conference on, 1999, pp. 244–251.
10. J. D. Madden and I. W. Hunter, "Three-dimensional microfabrication by localized electrochemical deposition," Microelectromechanical Systems, Journal of, vol. 5, no. 1, pp. 24–32, 1996.
11. S. Barcikowski, M. HUSTEDT, and B. CHICHKOV, "Nanocomposite manufacturing using ultrashort-pulsed laser ablation in solvents and monomers," Polimery, vol. 53, no. 9, pp. 657–662, 2008.
12. Z. B. Sun, X. Z. Dong, S. Nakanishi, W. Q. Chen, X. M. Duan, and S. Kawata, "Log-pile photonic crystal of CdS–polymer nanocomposites fabricated by combination of two-photon polymerization and in situ synthesis," Applied Physics A: Materials Science & Processing, vol. 86, no. 4, pp. 427–431, 2007.
13. X. -Z. D. Z. -B. Sun, "Log-pile photonic crystal of CdS–polymer nanocomposites fabricated by combination of two-photon polymerization and in situ synthesis," vol. 86, no. 4, pp. 427–431, 2007.
14. C. K. Malek and V. Saile, "Applications of LIGA technology to precision manufacturing of high-aspect-ratio micro-components and-systems: a review," Microelectronics Journal, vol. 35, no. 2, pp. 131–143, 2004.
15. R. K. Kupka, F. Bouamrane, C. Cremers, and S. Megtert, "Microfabrication: LIGA-X and applications," Applied Surface Science, vol. 164, no. 1–4, pp. 97–110, Sep. 2000.
16. R. Kondo, S. Takimoto, K. Suzuki, and S. Sugiyama, "High aspect ratio electrostatic micro actuators using LIGA process," Microsystem technologies, vol. 6, no. 6, pp. 218–221, 2000.
17. L. Hong and T. Pan, "PhotopatternableSuperhydrophobicNanocomposites for Microfabrication," Journal of Microelectromechanical Systems, vol. 19,

no. 2, pp. 246–253, Apr. 2010.

18. X. Zhang and C. Sun, "Experimental and numerical investigations on microstereolithography of ceramics," Journal of Applied Physics, vol. 92, no. 8, p. 4796, Jan. 2002.
19. J. Marqués-Hueso, R. Abargues, J. L. Valdés, and J. P. Martínez-Pastor, "Ag and Au/DNQ-novolacnanocompositespatternable by ultraviolet lithography: a fast route to plasmonic sensor microfabrication," J. Mater. Chem., vol. 20, no. 35, pp. 7436–7443, 2010.
20. M. Ouyang, R. Bai, Y. Xu, C. Zhang, C. A. Ma, M. Wang, and H. Z. Chen, "Fabrication of polypyrrole/TiO2 nanocomposite via electrochemical process and its photoconductivity," Transactions of Nonferrous Metals Society of China, vol. 19, no. 6, pp. 1572–1577, 2009.
21. D. Z. Chen, S. Lao, J. H. Koo, M. Londa, and Z. Alabdullatif, "Powder Processing and Properties Characterization of Polyamide 11-Graphene anocomposites for Selective Laser Sintering," in Proc. 2010 solid freeform fabrication symposium, Austin, TX, August, 2010, pp. 2–4.
22. M. Goeppert-Mayer, "ON ELEMENTARY ACTS WITH TWO QUANTUM JUMPS.," May 1967.
23. T. Baldacchini, "Acrylic-based resin with favorable properties for three-dimensional two-photon polymerization," Journal of Applied Physics, vol. 95, no. 11, p. 6072, Jan. 2004.
24. K. -S. Lee, D. -Y. Yang, S. H. Park, and R. H. Kim, "Recent developments in the use of two-photon polymerization in precise 2D and 3D microfabrications," Polymers for Advanced Technologies, vol. 17, no. 2, pp. 72–82, 2006.
25. M. Alubaidy, K. Venkatakrishnan, and B. Tan, "Fabrication of a reinforced polymer microstructure using femtosecond laser material processing," J. Micromech. Microeng., vol. 20, no. 5, p. 055012, May 2010.
26. M. Alubaidy, B. Tan, A. Mahmood, and K. Venkatakrishnan, "Nanofiber Plasmon Enhancement of Two-Photon Polymerization Induced by Femtosecond Laser," J. Nanotechnol. Eng. Med., vol. 1, no. 4, pp. 041015–041015, Nov. 2010.
27. B. Gu and W. Ji, "Two-step four-photon absorption," Opt Express, vol. 16, no. 14, pp. 10208–10213, Jul. 2008.
28. W. c. Oliver and G. m. Pharr, "An improved technique for determining hardness and elastic modulus using load and displacement sensing indentation experiments," Journal of Materials Research, vol. 7, no. 06, pp. 1564–1583, 1992.
29. B. Bhushan and X. Li, "Micromechanical and tribological characterization of doped single-crystal silicon and polysilicon films for microelectromechanical systems devices," Journal of Materials Research, vol. 12, no. 01, pp. 54–63, 1997.
30. S. Carusotto, G. Fornaca, and E. Polacco, "Multiphoton Absorption and Coherence," Phys. Rev., vol. 165, no. 5, pp. 1391–1398, Jan. 1968.
31. M. Alubaidy, B. Tan, A. Mahmood, and K. Venkatakrishnan, "Mechanical

Property Enhancement of Nanocomposite Microstructures Generated by Two Photon Polymerization," J. Nanotechnol. Eng. Med., vol. 1, no. 4, pp. 041016–041016, Nov. 2010.

32. P. K. Mallick, Fiber-Reinforced Composites: Materials, Manufacturing, and Design. M. Dekker, 1993.

33. N. T. K. Thanh, Magnetic Nanoparticles: From Fabrication to Clinical Applications. CRC Press, 2012.

34. P. Kruus, M. O'Neill, and D. Robertson, "Ultrasonic initiation of polymerization," Ultrasonics, vol. 28, no. 5, pp. 304–309, Sep. 1990.

35. G. J. Lee, S. H. Lee, K. S. Ahn, and K. H. Kim, "Synthesis and characterization of soluble polypyrrole with improved electrical conductivity," Journal of Applied Polymer Science, vol. 84, no. 14, pp. 2583–2590, 2002.

36. J. R. Li, J. R. Xu, M. Q. Zhang, and M. Z. Rong, "Carbon black/polystyrene composites as candidates for gas sensing materials," Carbon, vol. 41, no. 12, pp. 2353–2360, 2003.

37. J. X. Li, M. Silverstein, A. Hiltner, and E. Baer, "The ductile-to-quasi-brittle transition of particulate-filled thermoplastic polyester," Journal of Applied Polymer Science, vol. 52, no. 2, pp. 255–267, 1994.

38. S. Loshaek, "Crosslinked polymers. II. Glass temperatures of copolymers of methyl methacrylate and glycol dimethacrylates," Journal of Polymer Science, vol. 15, no. 80, pp. 391–404, Feb. 1955.

39. K. Matyjaszewski, Cationic Polymerizations: Mechanisms, Synthesis & Applications. Taylor & Francis, 1996.

40. M. -A. Alubaidy, K. Venkatakrishnan, and B. Tan, "Synthesis of magnetic nanofibers using femtosecond laser material processing in air," Nanoscale Research Letters, vol. 6, no. 1, p. 375, May 2011.

41. M. Alubaidy, L. Soleymani, K. Venkatakrishnan, and B. Tan, "Femtosecond laser nanostructuring for femtosensitive DNA detection," Biosensors and Bioelectronics, vol. 33, no. 1, pp. 82–87, Mar. 2012.

42. A. S. Mahmood, K. Venkatakrishnan, B. Tan, and M. Alubiady, "Effect of laser parameters and assist gas on spectral response of silicon fibrous nanostructure," Journal of Applied Physics, vol. 108, no. 9, pp. 094327–094327–6, 2010.

43. Y. Zhao, J. Zhai, S. Tan, L. Wang, L. Jiang, and D. Zhu, "TiO2 micro/nano-composite structured electrodes for quasi-solid-state dye-sensitized solar cells," Nanotechnology, vol. 17, p. 2090, 2006.

44. J. Bomm, A. Büchtemann, A. J. Chatten, R. Bose, D. J. Farrell, N. L. A. Chan, Y. Xiao, L. H. Slooff, T. Meyer, A. Meyer, W. G. J. H. M. van Sark, and R. Koole, "Fabrication and full characterization of state-of-the-art quantum dot luminescent solar concentrators," Solar Energy Materials and Solar Cells, vol. 95, no. 8, pp. 2087–2094, Aug. 2011.

Chapter 14

CELLULOSIC FIBERS: ROLE OF MATRIX POLYSACCHARIDES IN STRUCTURE AND FUNCTION

Polina Mikshina, Tatyana Chernova, Svetlana Chemikosova, Nadezhda Ibragimova, Natalia Mokshina and Tatyana Gorshkova

Kazan Institute of Biochemistry and Biophysics, Kazan Scientific Centre, Russian Academy of Sciences, Kazan, Russia

INTRODUCTION

Cellulose, being the major cell wall component and the most abundant organic matter, produced by living organisms, is not uniformly distributed within plant tissues. There are numerous cells, like the parenchyma ones, which even at maturity have thin cell wall. Thick cell walls are characteristic for the tissues with mechanical function. Among those, there are cell walls, which contain several major components, and those, the predominant component of which is cellulose. The most pure natural cellulose is considered to be present in cotton seed hairs (sometimes erroneously called "cotton fibers")

- over 90% of cell wall [1]. Very close to this value is a special group of plant fibers - cellulosic or gelatinous fibers, the proportion of cellulose in which amounts for 85-90% [2,3]. The cell wall thickness in such fibers may reach 15 μm, as compared to 0.2 μm in cells with thin cell wall. So, the very significant portion of total plant cellulose may be concentrated within the gelatinous fibers, making them the important source for production of biofuels and bio-based products. An additional attractiveness of cellulosic fibers for such applications comes from the fact that gelatinous cell wall layers are devoid of lignin - the major hurdle in using plant biomass [1].

Cell wall of cellulosic fiber is of very special design, which provides unusual properties. Such fibers serve as a kind of plant "muscles" [4,5]. The revealing of mechanisms of the formation and function of cellulosic fibers is important for understanding the determinants of general plant architecture and can be useful in construction of new bio-based materials.

DEFINITION OF PLANT CELLULOSIC FIBERS

In terms of plant biology, a fiber is an individual cell of mechanical tissue (sclerenchyma), characterized by the extreme cell length and well developed secondary cell wall, architecture of which is the major determinant of fiber properties. Plant fibers are one of the most economically important raw material, used both for traditional and innovative technologies [6,7]. Due to the massive cell wall, fibers comprise a significant proportion of plant biomass, thus being the valuable source of bio-based products and biofuels. For plant itself fibers are very important for the general architecture and mechanical properties of certain organs.

The functional roles of fibers within the plant and their numerous commercial applications are largely based on the characteristics of their well-developed cell wall of considerable thickness. Fibers of different origin are not uniform in their structure and cell wall composition. The thick cell walls of fully differentiated fibers can be categorized into two broad types - the xylan and the gelatinous ones [2] (Figure 1).

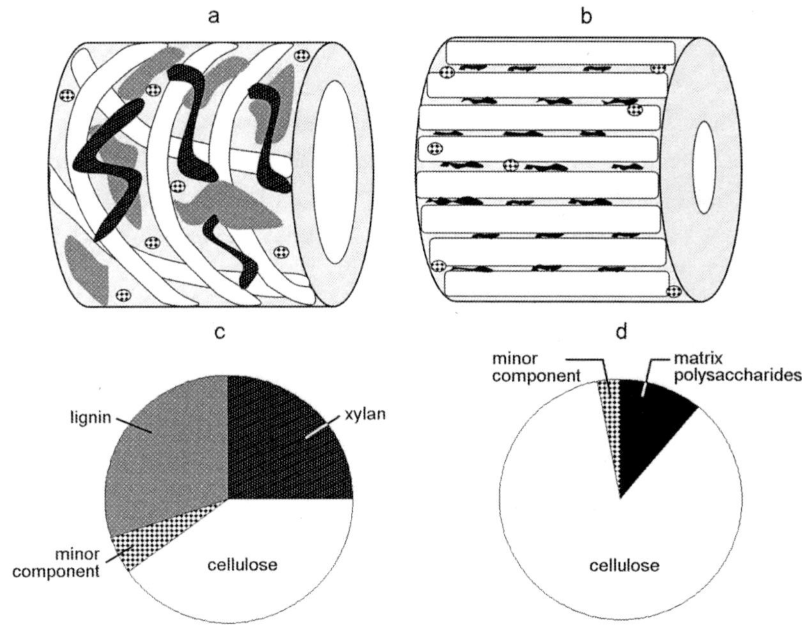

Figure 1. A scheme of structure (a, b) and content of the main components (c, d) in two types of the secondary cell walls: a, c – xylan type of cell wall, b, d – gelatinous type of cell wall.

A secondary cell wall of the xylan type is the most common one in various types of cells with secondary cell walls in land plants. The xylan type secondary cell walls are characterized by helical orientation of cellulose microfibrils, predominance of xylan in non-cellulose matrix, and high degree of lignification. The orientation of cellulose microfibrils may be significantly changed several times through the development of the xylan type secondary cell wall, leading to the formation of distinct layers, designated as S1, S2, and S3 layers (S from "secondary") in the order of deposition. Total thickness of the xylan type secondary cell walls is between 1 and 4 µm. They comprise the bulk of secondary cell walls in various types of wood cells, including vessels and wood parenchyma, and are also present in the bast fibers of some species, for example, jute and kenaf.

The second (gelatinous) type of thick cell wall is present only in fibers. It was firstly described by Th. Hartig at the end of the XIX century as a peculiar layer (G-layer) produced in the reaction wood of dicotyledonous plants (cited after [8]). The reason for such name came from the artefactual swelling of this layer in the cross-section due to the presence of certain components (e.g. alkali) of the solutions used to prepare the sample for microscopy. Fibers, which have developed G-layer of cell wall, get the name "gelatinous fibers". With modern techniques of sample preparation for microscopy, this layer doesn't look like a jelly. Moreover, G-layer was described as having exclusively high content of cellulose (up to 90%) with high degree of crystallinity [9], giving the reason for the alternative name – "cellulosic fibers". However, the justification for a term "gelatinous" was recently provided by description of the gel-like performance of G-layer upon drying (large shrinkage [8,10-12] and high rigidification [13]), and hydrogel type of structure, which has special characteristics of mesoporosity. G-layer has high content of mesopores (pore size between 2 nm and 50 nm); in tension wood fibers the pore surface areas may be more than 30 times higher than that in normal wood as was revealed by nitrogen adsorption technique [8, 14]. Mesoporosity was suggested as a new parameter for G-layer characterization [14].

Gelatinous cell-wall layer is deposited inward to the xylan type secondary wall layers; the degree of S-layer development in fibers with G-layer differs from well pronounced, like in tension wood [15], to barely detectable, as in flax [2]. Though not appropriately recognized, this type of fibers is widespread and is present in various organs of plants from many taxa [2,3]. Among others, phloem fibers of flax, hemp, and ramie, gelatinous fibers of tension wood, some fibers of bamboo and *Equisetum* belong to this group. Arabidopsis was shown to have the potential for gelatinous fiber formation [16], same as some other plant species where this type of fibers was not well known, like alfalfa [17]. Fibers able to form the gelatinous cell wall layer can originate from both primary and secondary meristems and be located within phloem or xylem [18].

Specific characteristics of the gelatinous layer of cell wall include: a) the overwhelming content of cellulose (80-90%); b) high crystallinity of cellulose; c) very low angle of cellulose microfibrils, which are laid almost parallel to the fiber's longitudinal axis throughout the whole layer; d) considerable thickness, which in some species can reach more than 15 µm; e) the absence of xylan, f) the absence of lignin, g) special composition of matrix polysaccharides, presented mostly by galactose-containing pectins with rhamnogalacturonan backbone; h) high water content, as compared to S-layers, i) mesoporosity, j) exclusive presence in fibers, and, as discussed below, k) contractile properties.

The very peculiar characteristic of fibers with the gelatinous cell wall is their contractile properties. Such fibers may serve to move the plant parts in space. For instance, the tension wood fibres contract longitudinally during differentiation and generate longitudinal tensile stresses of up to about 70 MPa [19], providing rightning force in the tilted tree. A high-tensile growth stress generated on the surface of the xylem in the tension wood region often becomes ten times as large as that in the normal wood region [20]. The ability of a plant organ to contract is proportional to the degree of the development of fibers with G-layers [21,22].

Plants do not possess animal-type muscles, which contract due to protein-protein interactions. However, they have a different mechanism, which has the ability to move even very heavy plant parts in space. This mechanism is specifically developed in cellulosic fibers. Their contractile properties are based on tension developing within the specially designed thick cell walls. The efficacy of such fibers is remarkable. Thus, the gelatinous fibers may to a certain extent be named as "cell-wall-based plant muscles", though they do not have the ability to relax and the time-scale of their contraction is very different from that of animal muscles.

The examples of contractile action of cellulosic fibers (Figure 2) include the restoration of young tree vertical position, if disturbed [15,23], deepening of geophyte shoot in the soil by contraction of roots in the course of adaptive reaction [24], shortening of aerial roots when they reach the soil to form the effective support to heavy branches [21]; correction of lateral branch angle and, in case of apical meristem death, upward bending of lateral branch in order to transform into dominant [25-27]. The cellulosic gelatinous fibers are especially developed in the plants exposed to high mechanical stresses due to high ratio between stem height and stem diameter (most of fiber crops, like flax, hemp, ramie, nettle) [2] – this gives the properties of spring to the plant stem, helping to restore vertical position when disturbed, for example by wind. Such plants, similar to plants with developed tension wood [28,29], also exhibit the pronounced negative gravitropism, being able to turn back to vertical position a long stem, if bent, far away from the growing apical stem region. Cellulosic fibers seem to be present in the spines of cacti [30], and are developed in peduncles helping to support heavy fruits, like in the sausage tree, fruits of which may weight up to 10 kg [31]. The gelatinous fibers were demonstrated to be widely involved in the twining of vines and the coiling of tendrils [32,33].

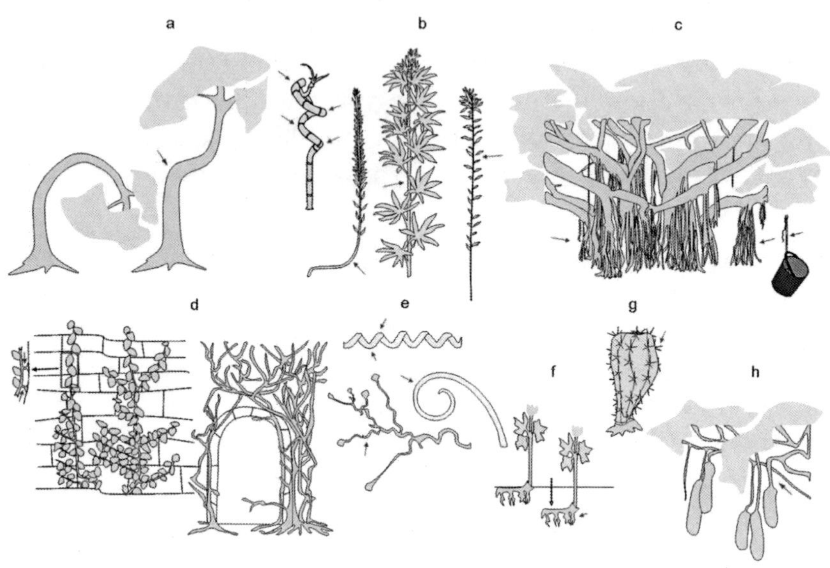

Figure 2. The examples of contractile action of cellulosic fibers: a – formation of tension wood during restoration of young tree vertical position, if disturbed, b – development of gelatinous fibers in the plants with high ratio between stem height and stem diameter, c – shortening of aerial roots after they reach the soil to form the effective support to the heavy branches (when rooted in pot, aerial roots may rise it up), d – development of cellulosic fibers in parts of stem due to which plant attaches to substrate, e – involvement of gelatinous fibers in the twining of vines, the coiling of tendrils and the expansion of climbing plants, f – deepening of geophyte shoot into the soil by contraction of roots in the course of adaptive reaction, g – presence of cellulosic fibers in the spines of cacti, h – development of gelatinous fibers in peduncles to help support heavy fruits.

CELLULOSE MICROFIBRILS IN GELATINOUS CELL WALL

The very specific characteristic of cellulose microfibrils in the gelatinous cell wall is their axial orientation, which is not observed in any other cell wall type of any other but fiber cell type. This is especially remarkable, if one remembers the total thickness of G-layers. The axial orientation of cellulose microfibrils throughout the gelatinous cell wall layer was known long ago [9,34] and was confirmed by several techniques, including microbeam X-ray diffraction [35], wide-angle X-ray scattering [36], and scanning Raman microscopy [24,37]. In accordance to that, cortical microtubules, which are considered to rule the microfibril orientation, are axially oriented during deposition of the gelatinous cell wall layers [38].

In the gelatinous fibers cellulose microfibrils are characterized by a higher degree of crystallinity and a larger size of crystalline regions (crystallites) as compared with most other plant tissues [39, 40]. The diameter of cellulose crystallites was measured in various species by several authors and though the absolute values might differ, the general conclusion was that in G-layers it was larger than in the S-layers [12,35,41-44]. The roentgen structural analysis showed that the diameter of cellulose crystal transverse sections in tension wood G-layer (6.5 nm) is markedly larger than in the neighboring S-layer of the xylan cell wall (about 3 nm), i.e., its section area is approximately fourfold higher [35]. Thus, cellulose microfibrils within the gelatinous layer exist in the form of aggregates. This allowed a supposition that individual cellulose microfibrils, each of which is formed by individual cellulose-synthesizing complex, so-called "rosette", in the gelatinous layer interact laterally [4]. Such lateral interaction is stimulated due to similar (axial) orientation of all microfibrils, the absence of lignin and of considerable amount of matrix polysaccharides, which separate microfibrils, like in S-layer. Despite the high degree of crystallinity of the cellulose, G-layer has a remarkable hygroscopicity and high water content [37,45].

Cellulose microfibrils in the gelatinous layer are under tension. This was proved by the increase in cellulose lattice spacing revealed by synchrotron radiation microdiffraction [46]. The visual demonstration of tension comes from shrinkage of G-layer along the cell longitudinal axis upon the release of tension, as observed by scanning microscopy of tension wood cross-sections [10].

For a long time, the gelatinous cell wall was believed to be composed of cellulose only [9]. Correspondingly, the ideas on tension origin were based on cellulose microfibril properties. Cellulose microfibrils themselves are virtually incontractible. So, the problem was how to get contraction, having incontractible basis of cellulose microfibrils. One of the possible solutions suggests that the contraction of the fibre is not caused by the G-layer directly, but by interaction of the G-layer with the surrounding S-layer. It was proposed that the origin of tension in cell wall of cellulosic fibers lays in the differential parameters of swelling of the S- and G-layers due to different orientation of cellulose microfibrils [36].

The cell walls with helicoidal orientation of microfibrils increase in length upon swelling, while the ones with axial orientation shrink [47]. The stress–strain-curves of cell walls show the influence of the cellulose microfibril orientation on the deformation behavior of plant tissues [48]. Within plant organism, such differences are indeed exploited in some mechanisms, for instance in opening of pine cones [49]. Fibers and sclereids located at the opposite sides of a cone scale have different angle between the long axis of the cell and the direction of cellulose microfibrils (MFA): it is high in sclerids and low in fibers. Correspondingly, the coefficient of hygroscopic expansion of fibers is significantly lower than that of sclerids. Due to that, the increase in relative humidity leads to the increase of the angle between the scale and the frame, leading to cone opening. Similar is the mechanism of wheat awn opening aimed to seed dispersal [48,50].

Similarly, the differences in swelling of S- and G-layers were suggested to explain the formation of tension in fibers of reaction wood [36]. The idea is based on the established fact that the enzymatic removal of the G-layer lead to the longitudinal extension and tangential shrinkage of tissues within the tension wood slice. It was proposed that in the living plant, a lateral swelling of the G-layer forced the surrounding S-layers to shrink in the axial direction.

Such forces may indeed be the part of the tension creation mechanisms in gelatinous fibers. But: 1) such system would be highly dependent on humidity, same as the openings of pine cones and wheat awns; 2) in some species, like flax or ramie, S-layer in the gelatinous fibers is poorly developed [2] and hardly may serve as a mechanical counterpart of very thick G-layer; 3) mesoporosity of the G-layer is not explained; 4) tension is argumented to be developed within the G-layer itself [10,51]. Finally, specific matrix polysaccharides, which were not considered in the above hypothesis, appear at the onset of G-layer formation.

MATRIX POLYSACCHARIDES IN CELLULOSIC FIBERS

The presence of a polymer within a certain cell wall layer is not easy to prove. The biochemical analysis in the majority of studies is usually performed without separation into different cell wall layers, which is rather hard to achieve. In most of the experiments on the analysis of the gelatinous fiber composition, primary cell wall was not detached from the secondary one, and the xylan layers were not separated from the gelatinous ones. Moreover, such analysis is often done on the samples, like wood, which contain complex mixture of various cell types (e.g. parenchyma, vessels, fibers) at different stages of the development. A significant amount of data concerning the composition and structure of the gelatinous type cell walls was obtained by the analysis of phloem fiber bundles, which extreme strength permited their mechanical or enzymatic separation from surrounding tissues. Although gelatinous layers predominate in such fiber cell walls, the primary cell wall and S-layer of the secondary wall are also present. That's why, for instance, polygalacturonic acid (PGA) or rhamnogalacturonan I (RG I), described in numerous papers on plant fibers (e.g. [52-55] can not be attributed to a certain cell wall layer. The presence of polymers just in the G-layers must be additionally proved.

To do so, several approaches can be used or, better, combined: a) isolation of the G-layers and the biochemical analysis of constituents; b) cytochemistry, including immunocytochemistry; c) the analysis of deposition dynamics: search for the marker monomer, sugar linkage type or other specific characteristic of a certain polysaccharide, the formation of which goes in parallel to the G-layer deposition; pulse-chase experiments with labeled precursors can be especially effective since they permit to exclude the background of previously synthesized polymers; d) tracing the transcription of the identified genes, involved in the metabolism of certain cell wall polysaccharide, in the course of the G-layer formation; e) detection within G-layer of the enzyme or enzymatic activity, involved in modification of matrix polysaccharide, by various types of staining. The best way to analyze the components of the gelatinous cell walls is isolation of the G-layers, like it was done for poplar tension wood [28,56]. To this end, thin tissue sections (20 μm) are prepared and treated with ultrasound; however, this procedure permits obtaining only small amount of the material.

Among neutral monosaccharides present in polymers of the isolated G-layers, rhamnose, arabinose, galactose, and xylose were found along with glucose [56]. The cell wall matrix polysaccharides, which were identified or suggested to be present in the G-layer include xyloglucan [56], arabinogalactan proteins [57], pectic galactan [58], and, probably some arabinans [30]. Usually these polymers are considered to be the components of the primary cell walls. The obtained data on their presence in the G-layer were recently summarized in the review [5], so here we consider them only briefly.

Arabinogalactan proteins (AGPs) are highly water soluble polymers, which consist of protein backbone and carbohydrate side chains of variable structure, which can comprise over 90% of the molecule. Glycan component of AGPs has various length chains of β-(1→4)-galactan and β-(1→6)-galactan units, often decorated by terminal arabinose residues and connected to each other by (1→3,1→6)-linked branch points, which are indicative of AGPs. Protein backbone may also vary in structure and is encoded by a large gene family, the several representatives of which are always detected among the most up-regulated upon the G-layer induction genes, both in tension wood [59-62] and in fiber crops [63-65]. Carbohydrate constituents of AGPs were detected within the G-layer by immunocytochemical [57,60], cytochemical (by staining with Yariv reagent) [58,66] and biochemical [56] approaches. AGPs, different both in carbohydrate and protein part of the molecules, are present in many, if not all, plant tissues, but their exact function is still unknown.

There is not much information about another possible constituents of the gelatinous fibers – the arabinans. They were reported to be the major cell wall matrix component of cellulosic fibers in cactus spines [30]. It is not clear if the arabinans are attached to the RG I backbones.

The most substantial evidence for the matrix cell wall polysaccharides of the G-layers was collected on xyloglucan. This cross-linking glycan is composed of a backbone, which is built similar to cellulose molecule as β-(1→4)-glucan. The side chain of xylose, which is sometimes additionally substituted by galactose and further - by fucose, are attached to the backbone. Xyloglucan is the major noncellulosic polysaccharide in the isolated G-layers of poplar tension wood; its content was assessed to be 10–15% of the cell wall mass [28,56,67]. The presence of xyloglucan was detected by several methods, including the biochemical analysis of the types of bonds between monosaccharides and immunocytochemistry. Moreover, the presence of xyloglucan endotransglycosylase, an enzyme providing for connection between the regions of two different xyloglucan molecules, was demonstrated in the G-layers of the secondary cell wall. Two main functions were suggested for xyloglucan in the secondary cell walls of tension wood fibers [67]. The first one is binding of the G-layer to the neighboring xylan layer because xyloglucan and xyloglucan endotransglycosylase are localized just at the boundary between these two layers, as was shown immunocytochemically. The second supposed function is the creation of tension – it will be considered in the next chapter.

One more component of the G-layers is pectic galactan, built as a very complex rhamnogalacturonan I with a high degree of branching and a varying structure of side chains, which are mainly composed of β-(1→4)-galactose [52,58,68]. The predominant monosaccharide in the polymer is galactose, which determines the polymer name as a galactan [69]. The side chains may include only one or two galactose residues; long chains of several tens galactose residues, likely branched side chains, which are not cleaved by galactanase; side chains, decorated with a single pentose, most likely arabinose or a galactose residue connected by other than β-(1→4) linkage [68,70]. This type of pectic galactan is fiber- and stage-specific, being present only in fibers, while forming G-layer [2,71].

Pectic galactan may be of the specific three-dimensional organization, the signs of which are revealed upon the treatment with specific glycanases. The hydrolysis of considerable part of galactose side chains of galactan as well as the partial degradation of its backbone do not change the total hydrodynamic volume, which determines the efficiency of elution from gel-filtration column, and the polymer elutes in the same part of profile, as before enzymatic treatments [68,72]. The unusual property of pectic galactan from the gelatinous fibers is the ability to form water-soluble associates, so that the charged backbone is located at the periphery of it, while the neutral side chains form the core zone (Mikshina, Gorshkova, in preparation).

The presence of the galactan within the gelatinous layer was confirmed by the analysis of the dynamics of its formation, which coincides with the G-layer deposition [71], by immunolocalization of the galactan side chains [71,73-75], by presence of tissue- and stage specific β-(1→4)-galactosidase, the substrate of which is the described galactan [75-77]. The gene of this galactosidase is highly upregulated at the onset of the G-layer formation [63, 78] and the activity is detected within fibers, forming gelatinous cell wall [64,75].

The complex galactans built mainly from β-(1→4)-galactose were found in tension wood fibers in 60-ties of the XXth century [79,80], though the linkage with the RG I backbone was not proved at that time. The content of galactose was even suggested as an indicator of the extent of the G-layer development [81]. However, these old data were actually put away for several decades due to the overwhelming notion that the G-layers were pure cellulosic, so that the published in 2008 paper describing the detection of rhamnogalacturonan I by cytochemical approaches in tension wood of several species was entitled "…gelatinous fibers contain more than just cellulose" [57].

MATRIX POLYMERS AS THE CAUSATIVE AGENT FOR CELLULOSE TENSION IN GELATINOUS CELL WALL

Presence of specific matrix polysaccharides within G-layer suggests their importance for function of cellulosic fibers, including tension creation to form contractile properties. Mellerowicz et al. [4] put forward an idea that matrix polysaccharides are entrapped by laterally interacting cellulose microfibrils. The presence of such entrapped polysaccharides between cellulose microfibrils limits their interaction and results in creation of tension, which underlies specific mechanical properties of cellulosic fibers (Figure 3).

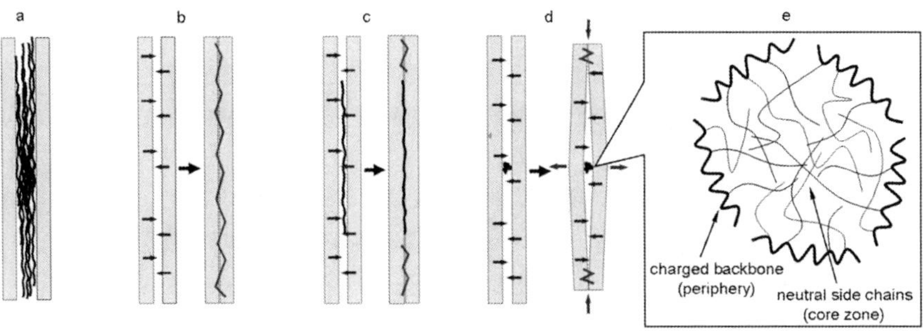

Figure 3. Possible ways of interaction between matrix polysaccharides and cellulose microfibrils in various types of cell walls: a – high content of matrix polysaccharides in xylan secondary cell wall prevents lateral interaction of cellulose microfibrils, b – microfibrils of the G-layer (gelatinous cell wall) with low content of matrix polysaccharides, cellulose microfibrils tend to lateral interactions, giving reason for higher degree of crystallinity and larger size of crystallites, c – theoretically, if matrix polysaccharides has high affinity to cellulose, being entrapped they won't cause much of tension, d – the most effective to provide longitudinal tensile stress in the cellulose microfibrils is compact polysaccharide of considerable size with low affinity to cellulose, e – a model of pectic galactan associates, in which negatively charged RG I backbone is at the periphery, and long galactose side chains form the core zone.

Originally, xyloglucan was proposed as the polymer entrapped by the laterally interacting cellulose microfibrils. Xyloglucan is, indeed, very important for the function of gelatinous fibers in tension wood of some species, like poplar. This is proved by the fact that in transgenic poplars with the expressed gene of fungal xyloglucanase, which decreased the content of xyloglucan, righting of stem basal regions in placed horizontally young plants was completely abolished, while the G-layer formation was not affected [29,82]. The expression of other endoglycanases, which decreased the levels of xylan or arabinogalactan, had no effect on the ability of transgenic plant to exhibit gravitropic reaction, restoring the stem vertical position.

However, the exact function of xyloglucan in tension wood fibers is still a matter of debate. Firstly, xyloglucan was detected in the G-layers only in limited plant species, it was never conclusively reported (though searched) to be present in thick secondary walls of cellulosic fibers in fiber crops, like flax, hemp, etc. Further evidence comes from the analysis of polysaccharides, strongly retained by cellulose microfibrils upon extraction: if entrapped between the interacting laterally cellulose microfibrils, a polymer should not be extracted by the conventional methods and should come out only after degradation of microfibrils by chemical or enzymatic means. However, due to high crystallinity of cellulose in the gelatinous layers, in natural form it is poorly degraded by specific enzymes [66] and thus, has to be first dissolved by corresponding chemicals.

To analyze the polysaccharides, which are especially strongly retained within cell wall, a special protocol was developed [83]. After removal of the extractable polysaccharides by chelators and concentrated alkali, the residual cell wall material was dissolved in solution of lithium chloride in N,N-dimethylacetamide and afterwards cellulose was precipitated by water. Such treatment turned natural cellulose I (with parallel orientation of individual cellulose chains) into cellulose II (with antiparallel orientation of individual cellulose chains) and made it completely degradable by purified cellulase. The matrix polysaccharides, which were present in the fraction, remained in polymeric form, making possible to separate them by gel-filtration for further analysis.

We have compared the composition of matrix polysaccharides, strongly retained by cellulose microfibrils, in fibers with different proportions of the secondary cell walls of xylan and gelatinous types (Figure 4). The polymers from fibers with only xylan type secondary cell wall eluted in the region below 30 kDa. The monosaccharide analysis and antibody binding indicated that the major component of this fraction was xylan. It is known that small proportion of matrix polymers, both in the primary and the secondary cell walls get entrapped by cellulose microfibrils in the process of their crystallization [84,85]. Some polygalacturonic acid was also present, which could be originated from the primary cell wall.

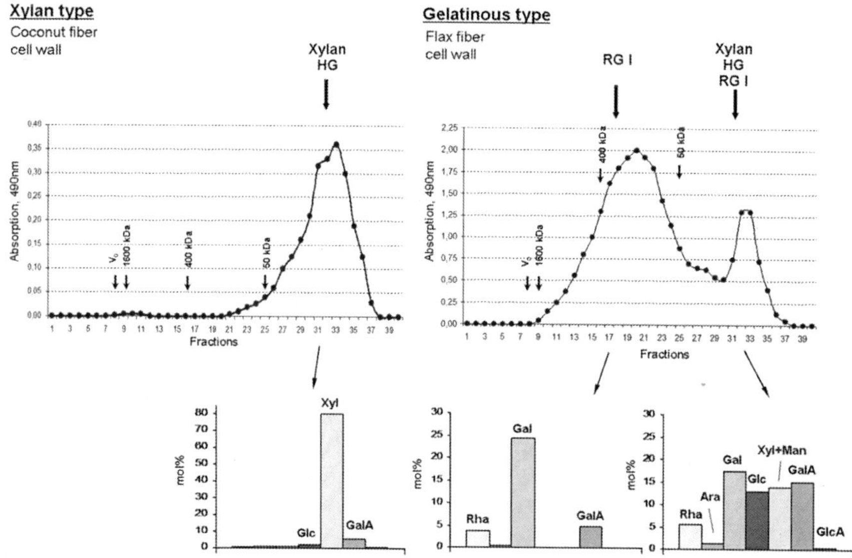

Figure 4. Elution profiles of polysaccharides strongly retained by cellulose microfibrils in the xylan and gelatinous cell walls and the relative monosaccharide content (mol%) of the main fractions of these polysaccharides.

In fibers with the G-layers, the major peak of matrix polysaccharide eluted between 100 and 400 kDa; its predominating component was pectic galactan. This galactan from flax fibers was characterized by various techniques, including ^{1}H and ^{13}C NMR and antibody binging [83]. The ratio between high and low molecular mass peaks on the elution profile depended on the proportion of the S- and G-layers within the fiber cell wall. Antibody to xyloglucan epitopes didn't bind any fraction on the elution profile.

The proportion of pectic polymers, which were strongly retained by cellulose microfibrils, from their total content in cell wall of the gelatinous type, could be much higher than that of xylan in the S-layers. Such selectivity in entrapping of certain polymers can not be explained by their affinity to cellulose as the charged pectic molecules are far less competitive, compared to xylan. The obtained data suggest the alternative mechanism of interaction between cellulose and pectic galactan, which is specifically developed in cell walls of the gelatinous type.

The above data suggest that in cellulosic fibers it is pectic galactan that is entrapped by laterally interacting cellulose microfibrils. This polymer, due to ability to form associates, can perfectly fit the proposed function in tension creation, as illustrated in Figure 3. In the xylan type of secondary cell wall (a), high content of matrix polysaccharides prevents the lateral interaction of cellulose microfibrils. At low content of matrix polysaccharides in G-layer (b), cellulose microfibrils tend to lateral interactions, giving reason for higher degree of crystallinity and larger size of crystallites. Matrix polysaccharides with high affinity to cellulose, if entrapped (c) won't cause much of tension. Most effective would be compact polysaccharide of considerable size with low affinity to cellulose (d). Associates of pectic galactan with RG I backbone may be a good choice of Nature for such purpose. They have compact structure of considerable volume, which has poor ability (due to charged surface) to interact with cellulose (e).

Additional arguments for the important role of pectic galactans in creation of tension come from the analysis of the course of the G-layer formation and of *in muro* modifications of matrix polymers, which was in detail performed on flax cellulosic fibers.

DYNAMICS OF THE G-LAYER FORMATION AND IN MURO MODIFICATIONS OF CELL WALL POLYMER

Formation of a cell wall layer is a complicated event. Partly it is based on the processes of polysaccharides' self-assembly in specific surroundings. Besides, the cell wall formation may involve modification of the interacting polysaccharides. The very illustrative example of the latter is the remodeling of the deposited G-layer in flax cellulosic fibers. In the dynamics of the G-layer formation two stages are clearly visualized at microscopic investigation of the flax fiber cell wall formation [71,86]. Under electron and/or light microscope, one can see that the inner part of the cell wall has a characteristic appearance of the loose structure where the electron dense parallel bands alternate with light regions; the outer part has much more homogenous structure (Figure 5). These two parts of the cell wall are designated as the Gn- and G-layers. During formation of the secondary cell wall, the thickness of outer layer gradually increases, while additional portions of the Gn-layer are added by the protoplast. This indicates that with time the Gn-layer is transformed into the G-layer.

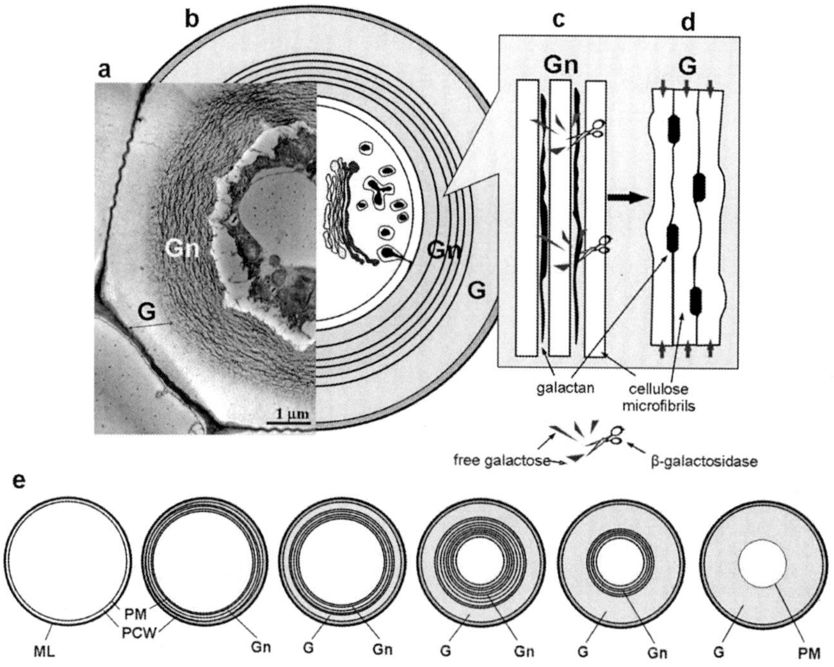

Figure 5. A model of the Gn-layer to the G-layer transformation in gelatinous fibers: a – electron microscopy of developing flax fiber cross-section; two layers (Gn and G) are obvious; b – scheme of developing flax fiber cross-section, showing a tissue-specific galactan delivered by specific Golgi vesicles to the developing Gn-layer; c – the nascent galactan is interspersed between cellulose microfibrils, preventing their association and maintaining the loosely packed morphology characteristic of the Gn-layer of secondary cell wall. During cell wall maturation, high molecular galactan partially digested by β-galactosidase, releasing free galactose; d – reducing of side chain length of galactan by galactosidase allows cellulose microfibrils to interact laterally, entrapping the galactan. Thus densely packed G-layer that is rich in crystalline cellulose is formed. The presence of entrapped galactan during lateral interactions of axially oriented microfibril causes longitudinal tensile stress in cellulose; e – dynamics of gelatinous layers deposition and remodeling in cellulosic fibers (left to right). ML – middle lamellae, PM – plasmalemma, PCW – primary cell wall, Gn – newly deposited gelatinous layer of secondary cell wall, G – mature gelatinous layer of secondary cell wall.

Transition from the Gn- to G-layer is coupled with changes in cellulose crystallinity. It is confirmed by the cytochemical analysis using the enzyme–gold complex, which showed that as distinct from G-layer, Gn-layer poorly bound cellobiohydrolase, the substrate of which is crystalline cellulose [58,87]. Besides, in pulse-chase experiments with $^{14}CO_2$ with intact flax plant, the dynamics of cellulose crystallization in fiber-enriched peels from all other analyzed samples. In roots, stem xylem, and stem apical part, which do not contain fibers with the gelatinous cell wall, the proportion of crystalline cellulose did not change during the entire experiment, while in fibers starting at the same level as in other tissues, it increased twice through the first day of chase and only later attained the plateau, which was at much higher level than in other tissues [88] (Figure 6). This indicated that crystallization of cellulose microfibrils in the G-layer was a biphasic process: the first stage occured right after the individual cellulose chain synthesis, similar to other plant tissues, while the second stage, which gave additional increase in crystallinity, occured *in muro* – within cell wall.

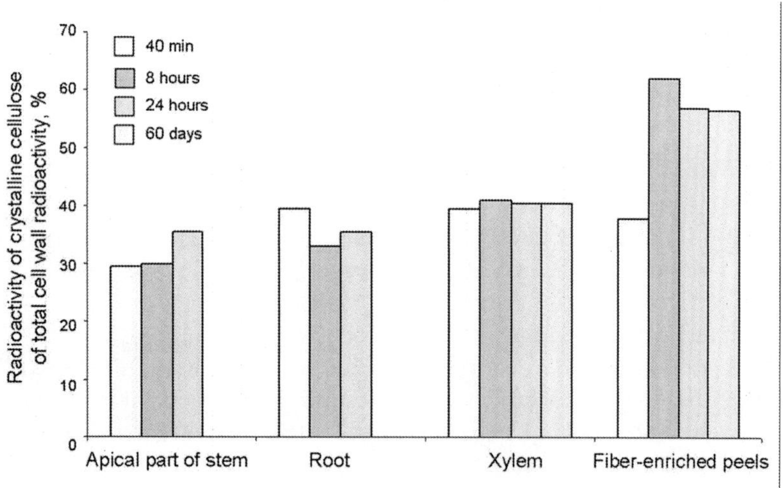

Figure 6. Radioactivity of crystalline cellulose of total cell wall radioactivity in the different part of flax plant after 40 min of photosynthesis with $^{14}CO_2$ (pulse) and during different periods of plant growth in the absence of a radioactive substrate (chase). Modified from data in [88].

The Gn- and G-layers differently bind not only cellobiohydrolase probe, but also the LM5 antibody, which is specific for β-(1→4)-galactan [89]. With LM5 the number of gold particles per area unit in the Gn-layer was fivefold higher than in the G-layer [71]. So, the reverse pattern was observed with binding the probes for cellulose crystallinity and for pectic galactan. Keeping in mind that antibody binding depends not only on the presence of the epitope but also on its availability, we consider it possible to suppose that changes in the degree of cellulose crystallization were related to *in muro*modification of tissue-specific galactan. An additional argument for such suggestion is a disappearance in the G-layer of dark bands, which are produced in the Gn-layer at galactan secretion by the Golgi apparatus and are well distinguished under electron microscope [71,74].

The pectic galactans are subjected to intensive *in muro* modifications. The investigation of galactan metabolism using the pulse-chase approach [2] confirmed that this polymer is synthesized in the Golgi apparatus, secreted outside the plasma membrane, and interacts with cellulose microfibrils. Flax fibers, while forming the secondary cell walls, have a peculiar mechanism of polysaccharide secretion. Golgi-derived vesicles first accumulate in the cytoplasm and only later fuse with the plasma membrane to give their contents to the apoplast [74]. These Golgi derivatives as well as the layers of the secondary cell wall, especially the inner "striated" layer, bind the LM5 antibody, indicating that all these structures contain galactan. It suggests that this peculiar type of galactan secretion permitting for filling large spaces of the periplasm facilitates the contact between galactan and cellulose microfibrils when they are in the process of assembly and may be necessary for preventing lateral interaction of cellulose microfibrils right at their deposition. Such a mechanism of secretion allows to accumulate a sufficient amount of the nascent pectic galactan before it is incorporated into the cell wall. This nascent form of the pectic galactan can be collected from the tissue homogenization buffer and compared to the polymer strongly retained by cellulose microfibrils. The composition and structure of these polysaccharides together with tracing in pulse-chase experiments [68-70,83] permit to consider the entrapped by cellulose microfibrils galactan as a derivative of the nascent galactan. The comparison of these polymers revealed the following differences:

the nascent polysaccharide elutes at gel-filtration as having higher molecular mass (in the 700-2000 kDa region) and has higher degree of branching and longer side chains, as compared to cell wall galactan [83].

The detected differences between the nascent and entrapped galactans suggested that they might be the result of *in muro* galactan modification by the enzyme cleaving off a part of the galactan side chains. Indeed, the histochemical staining of stems and hypocotyls with corresponding chromogenic or fluorogenic substrates shows β-galactosidase activity to be localized to developing fibers [64,76]. The gene of β-galactosidase is among the most up-regulated ones upon induction of the G-layer formation [63,78]. The substantial amounts of free galactose, which is the product of β-galactosidase action is present specifically in fibers forming gelatinous cell wall [76].

Shortening of the galactan side chains permits microfibril lateral interaction, due to which an additional portion of galactan is captured by them. The necessity of pectic galactan modification with the participation of β-galactosidase for the remodeling of cell wall supramolecular structure and transformation of the Gn-layer into mature the G-layer was demonstrated [75]. The role of fiber-specific β-galactosidase in providing the particular mechanical properties of gelatinous fibers was confirmed with transgenic flax plants (reduced galactan modifications – less mechanical strength) [75]. Antibodies raised to fiber-specific β-galactosidase of flax, revealed similar protein in the G-layers of cellulosic fibers in other plants (poplar tension wood fibers and both primary and the secondary phloem fibers of hemp), indicating that the process of the G-layer remodeling may be similar in fibers of different origin [77].

Thus, in the last several years the views on matrix polysaccharides of the gelatinous cell walls have changed dramatically: from rejecting their presence – to ascribing the major role to them in the development and function of cellulosic fibers.

CONCLUSIONS AND PERSPECTIVES FOR FUTURE RESEARCH

Summary of our ideas on the cell wall design of cellulosic fibers and the origin of their contractile properties include the following statements, based on the considered in the current review literature data and our own results:

Tension is caused due to lateral interaction of cellulose microfibrils and entrapment of matrix polysaccharides.

Lateral interaction is possible because of very high cellulose content, absence of xylan and lignin.

Similar axial orientation of all cellulose microfibrils in thick G-layer helps to cumulate tension of individual microfibrils and to develop it in the necessary direction. The effect is increased due to extreme length of fiber cells.

The entrapped polysaccharide - complex rhamnogalacturonan I with galactan side chains of specific structure and distribution, which is able to form water-soluble associates.

Entrapment of such associates leads to increased mesoporosity and to the development of cellulose microfibril tension.

High hydroscopic capacity of RG I helps to keep water in the G-layer.

Conditions for lateral interaction of cellulose microfibrils may be provided by *in muro* modification of deposited polysaccharide by fiber-specific galactosidase.

Additional important factors may be the interaction of the G-layer with the S-layer through the action of xyloglucan-modifying enzyme, the activity of which is mainly detected at the boundary between layers, and/or different deformation behavior of the S- and the G-layers upon swelling due to different orientation of cellulose microfibrils.

Cellulosic fibers are the example of very peculiar cell wall type. Its formation includes significant reprogramming of synthesis and secretion of matrix polysaccharides, reorientation of cellulose microfibrils, active remodeling of the deposited cell wall layers, specific inter- and intra-molecular interactions between cell wall polymers. The study of these processes may give additional clues for general understanding of the plant cell wall formation, which still belongs to the most enigmatic biological processes. Of special interest is the investigation of specific three-dimensional organization of pectic galactans from cellulosic fibers in order to elucidate the largely unknown principles of supramolecular structure of complex polysaccharides. Comparison of the gelatinous cell wall formation in fibers of various organs may help to figure out the biological determinants of plant fiber yield and quality in order to improve the characteristics of plant biomass for effective conversion into biofuels and bio-based products.

ACKNOWLEDGEMENT

This work was partially supported by the Russian Foundation for Basic Research (project no. 11-04-01602), and the Program of State Support of Leading Scientific Schools (project no. 825.2012.4, 12-04-31418).

REFERENCES

1. P. Albersheim, A. Darvill, K. Roberts, R. Sederoff, A. Staehelin, Plant Cell Walls. New York: Garland Science; 2011
2. T. A. Gorshkova, O. P. Gurjanov, P. V. Mikshina, N. N. Ibragimova, N. E. Mokshina, V. V. Salnikov, M. V. Ageeva, S. I. Amenitskii, T. E. Chernova, S. B. Chemikosova, Specific Type of Secondary Cell Wall Formed by Plant FibersRussian Journal of Plant Physiology201057328
3. T. Gorshkova, N. Brutch, B. Chabbert, M. Deyholos, T. Hayashi, S. Lev-yadun, E. J. Mellerowicz, C. Morvan, G. Neutelings, G. Pilate, Plant Fiber Formation: State of the Art, Recent and Expected Progress, and Open QuestionsCritical Reviews in Plant Sciences2012313201228
4. E. J. Mellerowicz, P. Immerzeel, T. Hayashi, Xyloglucan: the Molecular Muscle of TreesAnnals of Botany 2008102659
5. E. Mellerowicz, T. A. Gorshkova, Tensional Stress Generation in Gelatinous Fibres: a Review and Possible Mechanism Based on Cell-Wall Structure and Composition. Journal of Experimental Botany2012632551565
6. A. K. Bledzki, J. Gassan, Composites Reinforced with Cellulose Based FibresProgress in Polymer Science199924221
7. M. Jacob, S. Thomas, Biofibres and Biocomposites. Carbohydrate Polymers 200871343
8. B. Clair, J. Gril, Di Renzo F, Yamamoto H, Quignard F. Characterization of a Gel in the Cell Wall to Elucidate the Paradoxical Shrinkage of Tension Wood. Biomacromolecules20089494
9. P. H. Norberg, H. Meier, Physical and Chemical Properties of Gelatinous Layer in Tension Wood Fibers of Aspen (Populus Tremula L.). Holzforschung 196620174
10. B. Clair, B. Thibault, Shrinkage of the Gelatinous Layer of Poplar and Beech Tension

WoodIAWA Journal200122121

11. H. Yamamoto, K. Abe, Y. Arakawa, T. Okuyama, J. Gril, Role of the Gelatinous Layer (G-Layer) on the Origin of the Physical Properties of the Tension Wood of Acer SieboldianumJournal of Wood Science200551222
12. H. Yamamoto, J. Ruelle, Y. Arakawa, M. Yoshida, B. Clair, J. Gril, Origin of the Characteristic Hygro-Mechanical Properties of the Gelatinous Layer in Tension Wood from Kunugi Oak (Quercus Acutissima)Wood Science and Technology201044149
13. B. Clair, J. Ruelle, B. Thibaut, Relationship between Growth Stress, Mechanical-Physical Properties and Proportion of Fibre with Gelatinous Layer in Chestnut (Castanea Sativa Mill.)Holzforschung200357189
14. S. S. Chang, B. Clair, J. Ruelle, J. Beauchene, Di Renzo F, Quignard F, Zhao GJ, Yamamoto H, Gril J. Mesoporosity as a New Parameter for Understanding Tension Stress Generation in TreesJournal of Experimental Botany2009603023
15. E. J. Mellerowicz, M. Baucher, B. Sundberg, W. Boerjan, Unraveling Cell Wall Formation in the Woody Dicot Stem. Plant Molecular Biology 200147239
16. S. E. Wyatt, R. Sederoff, M. A. Flaishman, S. Lev-yadun, Arabidopsis Thaliana as a Model for Gelatinous Fiber FormationRussian Journal of Plant Physiology2010573363367
17. A. M. Patten, M. Jourdes, E. E. Brown, M-P. Laborie, L. B. Davin, N. G. Lewis, Reaction Tissue Formation and Stem Tensile Modulus Properties in Wild-Tipe and P-Coumarate-3- Hydroxylase Downregulated Lines of Alfalfa, Medicago Sativa (Fabaceae). American Journal of Botany 2007946912925
18. T. E. Chernova, T. A. Gorshkova, Biogenesis of Plant FibersRussian Journal of Developmental Biology200738221
19. T. Okuyama, H. Yamamoto, M. Yoshida, Y. Hattori, R. R. Archer, Growth Stresses in Tension Wood: Role of Microfibrils and LignificationAnnals of Forest Science 199451291
20. H. Yamamato, Role of the Gelatinous Layer on the Origin of the Physical Properties of the Tension Wood. Journal of Wood Science 200450197
21. M. H. Zimmermann, A. B. Wardrop, P. B. Tomlinson, Tension Wood in Aerial Roots of Ficus Benjamina L.Wood Science Technology 1968295
22. C. H. Fang, B. Clair, J. Gril, S. Q. Liu, Growth Stresses are Highly Controlled by the Amount of G-Layer in Poplar Tension WoodIAWA Journal200829237
23. G. Pilate, B. Chabbert, B. Cathala, A. Yoshinaga, J. C. Leple, F. Laurans, C. Lapierre, K. Ruel, Lignification and Tension Wood.Comptes Rendus Biologies2004327889
24. N. Schreiber, N. Gierlinger, N. Putz, P. Fratzl, C. Neinhuis, Burgert I. G-Fibres in Storage Roots of Trifolium Pretense (Fabaceae): Tensile Stress Generators for Contraction. The Plant Journal 201061854
25. J. B. Fisher, J. W. Stevenson, Occurrence of Reaction Wood in Branches of Dicotyledons and Its Role in Tree ArchitectureBotanical Gazette 198114282
26. Y. S. Hsu, S. J. Chen, C. M. Lee, L. L. Kuo-huang, Anatomical Characteristics of the Secondary Phloem in Branches of Zelkova Serrata MakinoBotanical Bulletin of Academia Sinica200546143

27. J. R. Hamilton, C. K. Thomas, K. L. Carvell, Tension Wood Formation Following Release of Upland Oak Advance Reproduction. Journal Wood and Fiber Science 1985173380390

28. swstmetapress.com/content/q70t6764831872r1/ (accessed 27 June 2007

29. T. Kaku, S. Serada, K. Baba, F. Tanaka, T. Hayashi, Proteomic Analysis of the G-Layer in Poplar Tension WoodJournal of Wood Science200955250

30. T. Hayashi, R. Kaida, T. Kaku, K. Baba, Loosening Xyloglucan Prevents Tensile Stress in Tree Stem Bending but Accelerates the Enzymatic Degradation of CelluloseRussian Journal of Plant Physiology201057316

31. M. R. Vignon, L. Heux, M. E. Malainine, M. Mahrouz, Arabinan-Cellulose Composite in Opuntia Ficus-Indica Prickly Pear Spines. Carbohydrate Research 20043391123131

32. P. Sivan, P. Mishra, K. S. Rao, Occurrence of Reaction Xylem in the Peduncle of Couroupita Guianensis and Kigelia PinnataIAWA Journal201031203

33. C. G. Meloche, J. P. Knox, K. C. Vaughn, A Cortical Band of Gelatinous Fibers Causes the Coiling of Redvine Tendrils: a Model Based upon Cytochemical and Immunocytochemical StudiesPlanta2007225485

34. A. J. Bowling, K. C. Vaughn, Gelatinous Fibers are Widespread in Coiling Tendrils and Twining VinesAmerican Journal of Botany200996719

35. H. E. Dadswell, A. B. Wardrop, The Structure and Properties of Tension WoodHolzforschung1955997

36. M. Müller, M. Burghammer, J. Sugiyama, Direct Investigation of the Structural Properties of Tension Wood Cellulose Microfibrils Using Microbeam X-Ray Fibre Diffraction. Holzforschung 200660474

37. L. Goswami, Dunlop JWC, Jungnikl K, Eder M, Gierlinger N, Coutand C, Jeronimidis G, Fratzl P, Burgert I. Stress Generation in the Tension Wood of Poplar is Based on the Lateral Swelling Power of the G-LayerThe Plant Journal200856531

38. N. Gierlinger, M. Schwanninger, Chemical Imaging of Poplar Wood Cell Walls by Confocal Raman MicroscopyPlant Physiology 20061401246

39. M. V. Ageeva, B. Petrovska, H. Kieft, Sal'nikov VV, Snegireva AV, van Dam JEG, van Veenendaal WLH, Emons AMC, Gorshkova TA, van Lammeren AAM. Intrusive Growth of Flax Phloem Fibers is of Intercalary TypePlanta 2005222565

40. R. J. Viëtor, R. H. Newman, M. A. Ha, D. C. Apperley, M. C. Jarvis, Conformational Features of Crystal-Surface Cellulose from Higher PlantsPlant Journal 200230721

41. A. Sturcova, I. His, D. C. Apperley, J. Sugiyama, M. C. Jarvis, Structural Details of Crystalline Cellulose from Higher Plants.Biomacromoles 200451333

42. R. Washusen, R. Evans, The Association Between Cellulose Crystallite Width and Tension Wood Occurrence in Eucalyptus GlobulusIAWA Journal200122235243

43. W. E. Hillis, R. Evans, R. Washusen, An Unusual Formation of Tension Wood in a Natural Forest Acacia sp.Holzforschung200458241

44. J. Ruelle, H. Yamamoto, B. Thibaut, Growth Stresses and Cellulose Structural Parameters in Tension and Normal Wood from Three Tropical Rainforest Angiosperms Species. Bioresources 20072235

45. M. Yamamoto, T. Saito, A. Isogai, M. Kurita, T. Kondo, T. Taniguchi, R. Kaida, K. Baba, T. Hayashi, Enlargement of Individual Cellulose Microfibrils in Transgenic Poplars Over Expressing Xyloglucanase. Journal of Wood Science 20115771

46. K. Abe, H. Yamamoto, The Influences of Boiling and Drying Treatments on the Behaviors of Tension Wood with Gelatinous Layers in Zelkova SerrataJournal of Wood Science2007535

47. B. Clair, T. Almeras, G. Pilate, D. Jullien, J. Sugiyama, C. Riekel, Maturation Stress Generation in Poplar Tension Wood Studied by Synchrotron Radiation Microdiffraction.Plant Physiology20101521650

48. I. Burgert, M. Eder, N. Gierlinger, P. Fratzl, Tensile and Compressive Stresses in Tracheids are Induced by Swelling Based on Geometrical Constraints of the Wood CellPlanta2007226981

49. I. Burgert, P. Fratzl, Plants Control the Properties and Actuation of Their Organs through the Orientation of Cellulose Fibrils in Their Cell Walls.Integrative and Comparative Biology20094969

50. C. Dawson, Vincent JFV, Rocca AM. How Pine Cones OpenNature1997668 EOF

51. R. Elbaum, S. Gorb, P. Fratzl, Structures in the Cell Wall that Enable Hygroscopic Movement of Wheat Awns.Journal of Structural Biology2008164101

52. B. Clair, T. Almeras, G. Pilate, D. Jullien, J. Sugiyama, C. Riekel, Maturation Stress Generation in Poplar Tension Wood Studied by Synchrotron Radiation Microdiffraction. Plant Physiology2011155562

53. E. A. Davis, C. Derouet, Herve du Penhoat C, Morvan C Isolation and N.M.R. Study of Pectins from Flax (Linum usitatissirnum L.). Carbohydrate Research 1990197205

54. C. Mooney, T. Stolle-smits, H. Schols, E. De Jong, Analysis of Retted and Non Retted Flax Fibres by Chemical and Enzymatic Means.Journal of Biotechnology200189205

55. C. Morvan, C. Andeme-onzighi, R. Girault, D. S. Himmelsbach, A. Driouich, D. E. Akin, Building Flax Fibres: More Than One Brick in the WallsPlant Physiology and Biochemistry 200341935

56. D. Cronier, B. Monties, B. Chabbert, Structure and Chemical Composition of Bast Fibers Isolated from Developing Hemp Stem.Journal of Agricultural and Food Chemistry2005538279

57. N. Nishikubo, T. Awano, A. Banasiak, V. Bourquin, F. Ibatullin, R. Funada, H. Brumer, T. T. Teeri, T. Hayashi, B. Sundberg, E. J. Mellerowicz, Xyloglucan Endotransglycosylase (XET) Functions in Gelatinous Layers of Tension Wood Fibers in Poplar- A Glimpse into the Mechanism of the Balancing Act of Trees. Plant and Cell Physiology 200748843

58. A. J. Bowling, K. C. Vaughn, Immunocytochemical Characterization of Tension Wood: Gelatinous Fibers Contain More Than Just CelluloseAmerican Journal of Botany200895655

59. T. Gorshkova, C. Morvan, Secondary Cell-Wall Assembly in Flax Phloem Fibers: Role of Galactans. Planta 2006223149

60. A. Dejardin, J-C. Leple, M-C. Lesage-descauses, G. Costa, G. Pilate, Expressed Sequence Tags from Poplar Wood Tissues- a Comparative Analysis from Multiple Libraries. Plant Biology 2004655

61. F. Lafarguette, J-C. Leple, A. Dejardin, F. Laurans, G. Costa, M-C. Lesage-descauses, G. Pilate, Poplar Genes Encoding Fasciclin-Like Arabinogalactan Proteins are Highly Expressed in Tension WoodNew Phytologist2004164107

62. S. Andersson-gunneras, E. J. Mellerowicz, J. Love, B. Segerman, Y. Ohmiya, P. M. Coutinho, P. Nilsson, B. Henrissat, T. Moritz, B. Sundberg, Biosynthesis of Cellulose-Enriched Tension Wood in Populus: Global Analysis of Transcripts and Metabolites Identifies Biochemical and Developmental Regulators in Secondary Wall Biosynthesis.The Plant Journal200645144

63. D. Qiu, I. W. Wilson, S. Gan, R. Washusen, G. F. Moran, S. G. Southerton, Gene Expression in Eucalyptus Branch Wood with Marked Variation in Cellulose Microfibril Orientation and Lacking G-LayersNew Phytology 200817994

64. M. J. Roach, M. K. Deyholos, Microarray Analysis of Flax (Linum Usitatissimum L.) Stems Identifies Transcripts Enriched in Fibre-Bearing Phloem TissuesMolecular Genetics and Genomics2007278149

65. M. J. Roach, M. K. Deyholos, Microarray Analysis of Developing Flax Hypocotyls Identifies Novel Transcripts Correlated with Specific Stages of Phloem Fibre DifferentiationAnnals of Botany 2008102317

66. N. Hobson, M. J. Roach, M. K. Deyholos, Gene Expression in Tension Wood and Bast Fibers. Russian Journal of Plant Physiology 201057339

67. R. Girault, I. His, C. Andeme-onzighi, A. Driouich, C. Morvan, Identification and Partial Characterization of Proteins and Proteoglycans Encrusting the Secondary Cell Walls of Flax Fibres.Planta2000211256

68. E. Mellerowicz, B. Sundberg, Wood Cell Walls: Biosynthesis, Developmental Dynamics and Their Implication for Wood Properties.Current Opinion in Plant Biology200811293

69. O. P. Gurjanov, T. A. Gorshkova, M. A. Kabel, H. A. Schols, van Dam JEG. MALDI-TOF MS Evidence for the Linking of Flax Bast Fibre Galactan to Rhamnogalacturonan BackboneCarbohydrate Polymers20076786

70. T. A. Gorshkova, S. E. Wyatt, V. V. Salnikov, D. M. Gibeaut, M. R. Ibragimov, V. V. Lozovaya, N. C. Carpita, Cell-Wall Polysaccharides of Developing Flax PlantsPlant. Physiology 19961102721729

71. P. V. Mikshina, O. P. Gurjanov, F. K. Mukhitova, A. A. Petrova, A. S. Shashkov, T. A. Gorshkova, Structural Details of Pectic Galactan from the Secondary Cell Walls of Flax (Linum Usitatissimum L.) Phloem FibresCarbohydrate Polymers201287853

72. T. A. Gorshkova, S. B. Chemikosova, V. V. Salnikov, N. V. Pavlencheva, O. P. Gurjanov, T. Stoll-smits, van Dam JEG. Occurrence of Cell-Specific Galactan is Coinciding with Bast Fibre Developmental Transition in FlaxIndustrial Crops and Products200419217

73. A. Gorshkova, P. V. Mikshina, N. N. Ibragimova, N. E. Mokshina, T. E. Chernova, O. P. Gurjanov, S. B. Chemikosova, Pectins in Secondary Cell Walls: Modifications during Cell Wall Assembly and Maturation. In: Schols HA, Visser RGF, Voragen AGJ. (eds) Pectins and Pectinases. Wageningen: Academic Publishers; 2009149164

74. M. Arend, A. Stinzing, C. Wind, K. Langer, A. Latz, P. Ache, J. Fromm, R. Hedrich, Polar-Localized Poplar K+ Channel Capable of Controlling Electrical Properties of Wood-Forming Cells. Planta 2005223140

75. V. V. Salnikov, M. V. Ageeva, T. A. Gorshkova, Homofusion of Golgi Secretory Vesicles in Flax Phloem Fibers during Formation of Gelatinous Secondary Cell Wall. Protoplasma2008233269

76. M. J. Roach, N. Y. Mokshina, A. V. Snegireva, A. Badhan, N. Hobson, M. K. Deyholos, T. A. Gorshkova, Development of Cellulosic Secondary Walls in Flax Fibers Requires ß-Galactosidase. Plant Physiology 20111561351

77. P. V. Mikshina, S. B. Chemikosova, N. E. Mokshina, N. N. Ibragimova, Gorshkova TA Free Galactose and Galactosidase Activity in the Course of Flax Fiber Development. Russian Journal of Plant Physiology 20095658

78. N. E. Mokshina, N. N. Ibragimova, V. V. Salnikov, S. I. Amenitskii, T. A. Gorshkova, Galactosidase of Plant Fibers with Gelatinous Cell Wall: Identification and LocalizationRussian Journal of Plant Physiology2012592246254

79. S. I. Amenitsky, A. V. Snegireva, T. A. Gorshkova, Genomics of plant fibers. Mutants with modified character of the fiber development. In: Gorshkova T. (ed.) Biogenesis of plant fibers. Moscow: Nauka; 2009189207

80. H. Meier, Studies on a Galactan from Tension Wood of Beech (Fagus Silvatica L.)Acta Chemica Scandinavica 1962162275

81. C. M. Kuo, T. E. Timell, Isolation and Characterization of a Galactan from Tension Wood of American beech (Fagus Grandifolia Ehrl.). Svensk Papperstid 196972703

82. K. Ruel, F. Barnoud, Goring DAI. Lamellation in the S2 Layer of Softwood Tracheids as Demonstrated by Scanning Transmission Electron MicroscopyWood Science and Technology1978124287291

83. K. Baba, Y. W. Park, T. Kaku, R. Kaida, M. Takeuchi, M. Yoshida, Y. Hosoo, Y. Ojio, T. Okuyama, T. Taniguchi, Y. Ohmiya, T. Kondo, Z. Shani, O. Shoseyov, T. Awano, S. Serada, N. Norioka, S. Norioka, T. Hayashi, Xyloglucan for Generating Tensile Stress to Bend Tree Stem.Molecular Plant20092893

84. O. P. Gurjanov, N. N. Ibragimova, O. I. Gnezdilov, T. A. Gorshkova, Polysaccharides, Tightly Bound to Cellulose in Cell Wall of Flax Bast Fibre: Isolation and IdentificationCarbohydrate Polymers200872719

85. T. Hayashi, Xyloglucans in the primary cell wallPlant Physiology. Plant Molecular Biology198940139

86. M. Pauly, P. Albersheim, A. Darvill, W. S. York, Molecular Domains of the Cellulose/ Xyloglucan Network in the Cell Walls of Higher Plants.Plant Journal 199920629

87. V. V. Salnikov, M. V. Ageeva, V. N. Yumashev, V. V. Lozovaya, The Ultrastructure of Bast Fibers. Russian Journal of Plant Physiology 199340458

88. C. Andeme-onzighi, R. Girault, I. His, C. Morvan, A. Driouich, Immunocytochemical Characterization of Early-Developing Flax Fiber Cell WallsProtoplasma2000213235

89. T. A. Gorshkova, S. B. Chemikosova, V. V. Lozovaya, N. C. Carpita, Turnover of Galactans and Other Cell Wall Polysaccharides during Development of Flax Plants. Plant Physiology 19971142723729

90. L. Jones, G. Seymour, J. P. Knox, Localization of Pectic Galactan in Tomato Cell Wall Using a Monoclonal Antibody Specific to (1?4)-ß-D-Galactan. Plant Physiology 19971131405

Citations

CHAPTER 1

Owais Anwar Golraa*, Jawad Tariqb, Nadeem Ehsanc and Ebtisam Mirzad Strategy for introducing 3D fiber reinforced composites weaving technology, http://dx.doi.org/10.1016/j.protcy.2012.02.046.

CHAPTER 2

Pragnesh N. Dave, Nikul N. Patel, Studies on Interacting Blends of Acrylated Epoxy Resin Based Poly(Ester-Amide)s and Vinyl Ester Resin, doi:10.4236/msa.2011.27106 Published Online July 2011.

CHAPTER 3

Milan Žmindák, Martin Dudinský. Computational Modelling of Composite Materials Reinforced by Glass Fibers, doi: 10.1016/j.proeng.2012.09.573.

CHAPTER 4

Ivan Pelivanov, Takashi Buma, Jinjun Xia, Chen-Wei Wei, Matthew O'Donnell NDT of Fiber-Reinforced Composites With A New Fiber-Optic Pump-Probe Laser-Ultrasound System dx.doi.org/10.1016/j.pacs.2014.01.001.

CHAPTER 5

Fengge Gao, Clay/polymer composites: the story, Materials Today, Volume 7, Issue 11, November 2004, Pages 50-55, ISSN 1369-7021, http://dx.doi.org/10.1016/S1369-7021(04)00509-7.

CHAPTER 6

O. D. Samuel, S. Agbo, T. A. Adekanye Leave a comment, Assessing Mechanical Properties of Natural Fibre Reinforced Composites for Engineering Applications , ISSN Online: 2327-4085.

CHAPTER 7

Influence of the composite surface structure on the peel strength of metallized carbon fibre-reinforced epoxy E. Njuhovic, A. Witta, M. Kempf, F. Wolff-Fabris, S. Glöde, V. Altstädt, http://dx.doi.org/10.1016/j.surfcoat.2013.05.025.

CHAPTER 8

Amal A.M. Badawy, Impact behavior of glass fibers reinforced composite laminates at different temperaturesdx.doi.org/10.1016/j.asej.2012.01.001.

CHAPTER 9

E.Z.Li, W.L.Guo, H.D.Wang, B.S.Xu,X.T.Liu, Research on Tribological Behavior of PEEK and Glass Fiber Reinforced PEEK Composite, doi: 10.1016/j.phpro.2013.11.071

CHAPTER 10

T. Alomayri, I.M. Low, Synthesis and characterization of mechanical properties in cotton fiber-reinforced geopolymer composites, http://dx.doi.org/10.1016/j.jascer.2013.01.002

CHAPTER 11

Nishiwaki, T.; Kwon, S.; Homma, D.; Yamada, M.; Mihashi, H. Self-Healing Capability of Fiber-Reinforced Cementitious Composites for Recovery of Watertightness and Mechanical Properties. Materials 2014, 7, 2141-2154. doi:10.3390/ma7032141.

CHAPTER 12

Vincenzo Giamundo, Gian Piero Lignola, Andrea Prota and Gaetano Manfredi, Nonlinear Analyses of Adobe Masonry Walls Reinforced with Fiberglass Mesh, DOI: 10.3390/polym6020464.

CHAPTER 13

A. Alubaidy, K. Venkatakrishnan and B. Tan (2013). Nanofibers Reinforced Polymer Composite Microstructures, Advances in Nanofibers, Dr. Russel Maguire (Ed.), ISBN: 978-953-51-1209-9, InTech, DOI: 10.5772/57101.

CHAPTER 14

Polina Mikshina, Tatyana Chernova, Svetlana Chemikosova, Nadezhda Ibragimova, Natalia Mokshina and Tatyana Gorshkova (2013). Cellulosic Fibers: Role of Matrix Polysaccharides in Structure and Function, Cellulose - Fundamental Aspects, Dr. Theo G.M. Van De Ven (Ed.), ISBN: 978-953-51-1183-2, InTech, DOI: 10.5772/51941.

INDEX

V

W

Y